# Lecture Notes in Mechanical Engineering

**Lecture Notes in Mechanical Engineering (LNME)** publishes the latest developments in Mechanical Engineering—quickly, informally and with high quality. Original research reported in proceedings and post-proceedings represents the core of LNME. Volumes published in LNME embrace all aspects, subfields and new challenges of mechanical engineering. Topics in the series include:

- Engineering Design
- Machinery and Machine Elements
- Mechanical Structures and Stress Analysis
- Automotive Engineering
- Engine Technology
- Aerospace Technology and Astronautics
- Nanotechnology and Microengineering
- Control, Robotics, Mechatronics
- MEMS
- Theoretical and Applied Mechanics
- Dynamical Systems, Control
- Fluid Mechanics
- Engineering Thermodynamics, Heat and Mass Transfer
- Manufacturing
- Precision Engineering, Instrumentation, Measurement
- Materials Engineering
- Tribology and Surface Technology

To submit a proposal or request further information, please contact the Springer Editor of your location:

**China:** Ms. Ella Zhang at ella.zhang@springer.com
**India:** Priya Vyas at priya.vyas@springer.com
**Rest of Asia, Australia, New Zealand:** Swati Meherishi at swati.meherishi@springer.com
**All other countries:** Dr. Leontina Di Cecco at Leontina.dicecco@springer.com

To submit a proposal for a monograph, please check our Springer Tracts in Mechanical Engineering at http://www.springer.com/series/11693 or contact Leontina.dicecco@springer.com

**Indexed by SCOPUS. All books published in the series are submitted for consideration in Web of Science.**

More information about this series at http://www.springer.com/series/11236

P. Srinivasa Pai · V. Krishnaraj
Editors

# Sustainable Machining Strategies for Better Performance

## Select Proceedings of SMSBP 2020

Springer

*Editors*
P. Srinivasa Pai
Department of Mechanical Engineering
NMAM Institute of Technology
Nitte, India

V. Krishnaraj
Department of Production Engineering
PSG College of Technology
Coimbatore, India

ISSN 2195-4356 ISSN 2195-4364 (electronic)
Lecture Notes in Mechanical Engineering
ISBN 978-981-16-2277-9 ISBN 978-981-16-2278-6 (eBook)
https://doi.org/10.1007/978-981-16-2278-6



This Springer imprint is published by the registered company Springer Nature Singapore Pte Ltd.
The registered company address is: 152 Beach Road, #21-01/04 Gateway East, Singapore 189721, Singapore

# NCSMSBP 2020 Organizing Committee

**Patron**

Sri N. Vinaya Hegde, Chancellor, Nitte Deemed to be University, President, Nitte Education Trust

**Chairman**

Dr. Niranjan N. Chiplunkar, Principal

**Organizing Chairman**

Dr. Shashikantha Karinka, Professor and Head

**Organizing Secretary**

Dr. P. Srinivasa Pai, Professor

**Conveners**

Dr. Grynal D'Mello, Assistant Professor
Mr. Bhaskara P. Achar, Assistant Professor

**Program Committee**

Dr. B. R. Shrinivasa Rao, Vice Principal and COE
Dr. I. R. Mithanthaya, Vice Principal and Dean (Academics)
Shri Yogesh Hegde, Registrar
Dr. Sudesh Bekal, Dean (R&D)
Dr. K. Subrahmanya Bhat, Dean (Student Welfare)
Dr. K. Rajesh Shetty, Dean (Admission and Alumni Affairs)
Prof. A. N. Parameswaran, Director, IIC
Dr. Srinikethan, Director (Technical Research)
Dr. Muralidhara, Professor and P.G. Coordinator

Dr. Narasimha Marakala, Professor
Dr. K. Srinath Shetty, Resident Engineer
Dr. Nithin Kumar, Assistant Professor
Dr. Rashmi P. Shetty, Assistant Professor
Mr. A. Adarsh Rai, Assistant Professor
Mr. J. S. Vishwanatha, Assistant Professor

**National Advisory Board**

Dr. P. V. Rao, Dean (Planning), Professor, Department of Mechanical Engineering, IIT Delhi
Dr. D. Chakradhar, Assistant Professor, Department of Mechanical Engineering, IIT Palakkad
Dr. Shrikanth Rao, Professor and Head, Department of Mechanical Engineering, NITK Surathkal
Dr. Dinesh Singh G. Thakur, Professor, Department of Mechanical Engineering, DIAT, Pune
Dr. V. Krishnaraj, Professor, Department of Production Engineering, PSGCT, Coimbatore
Dr. G. L. Samuel, Professor, Department of Mechanical Engineering, IIT Madras
Dr. Vishal Sharma, Professor, Department of IPE, BANIT Jalandhar
Shri. S. Gopinath, Former ED, BHEL, Trichy, Advisor, NMAMIT, Nitte
Dr. V. Ravichandran, Former GM, BHEL, Trichy, Adjunct Professor, NMAMIT, Nitte
Dr. Udaya Bhat, Professor, Department of Metallurgical and Materials Engineering, NITK Surathkal

**Technical Review Committee**

Dr. B. M. Rajprakash, Professor, Department of Mechanical Engineering, UVCE, Bengaluru
Dr. Y. S. Varadarajan, Professor, Department of Industrial and Production Engineering, NIE Mysore
Dr. V. Krishnaraj, Professor, Department of Industrial and Production Engineering, PSGCE, Coimbatore
Dr. S. S. Sharma, Professor, Department of Mechanical and Manufacturing Engineering, Manipal Institute of Technology, Manipal
Dr. Muralidhara, Professor, Department of Mechanical Engineering, NMAMIT, Nitte
Dr. K. V. Arun, Professor and Head, Department of Mechanical Engineering, GEC, Haveri
Dr. Chakradhar, Professor, Department of Mechanical Engineering, UVCE, Bengaluru
Dr. M. Sudheer, Professor, Department of Mechanical Engineering, St. Joseph Engineering College, Mangaluru
Dr. G. S. Vijay, Associate Professor, Department of Mechanical and Manufacturing Engineering, Manipal Institute of Technology, Manipal

Dr. Gururaj Bolar, Assistant Professor, Department of Mechanical and Manufacturing Engineering, Manipal Institute of Technology, Manipal
Dr. B. Sachin, Assistant Professor, Department of Mechanical Engineering, NMIT, Bengaluru
Dr. Charitha M. Rao, Assistant Professor, Department of Mechanical Engineering, Jain College of Engineering, Bengaluru
Dr. Vijeesh Vijayan, Associate Professor, Department of Mechanical Engineering, NMAMIT, Nitte
Dr. Ajith M. Hebbale, Associate Professor, Department of Mechanical Engineering, NMAMIT, Nitte
Dr. Grynal D'mello, Assistant Professor, Department of Mechanical Engineering, NMAMIT, Nitte

# Preface

Machining is a widely used manufacturing operation, which is based on subtractive mode. It requires a work piece material from which material needs to be removed using a cutting tool, which is harder. There can be single-point cutting tool or multi-point cutting tools used during machining, and the corresponding operations are turning, shaping and drilling, milling, grinding, etc. Machining operations result in the generation of chips, heat, tool wear, generation of cutting forces, vibrations, etc. In the last 100 years or so, significant research and developments are happening with regard to improving the productivity of these machining operations.

To remove the chips generated, reduce heat, tool wear, etc., coolants are being used, which provide cooling and lubrication effects. Machining industry is one such industry which consumes maximum energy and generates significant waste. In this regard, efforts are being made to reduce the use of coolants/lubricants, produce better quality tools, tool coatings, machine tools, etc., since last several years. The focus on "sustainability" in all aspects of life has also been brought into machining. Efforts are on to adopt sustainable machining practices in order to reduce waste, reduce environmental impact and conserve energy. Researchers are looking into various sustainability issues in machining in terms of technology, material and energy. In the Indian context, the availability of skilled manpower has led to use of conventional machine tools, without going in for CNC machine tools. Adoption of CNC technology helps in effectively bringing in "Life Cycle Assessment." The use of conventional flood lubrication is being reduced, and it is replaced either with "dry" machining or techniques like minimum quantity lubrication (MQL), cryogenic cooling and hybrid strategies which are being used.

In this context, there is a need to understand the sustainability issues in machining and be aware of the recent developments. This conference hopes to discuss major issues related to bringing in sustainability into machining processes, particularly with emphasis on difficult-to-machine materials, which create a lot of problems during machining. This includes tool wear, high cutting forces, high temperature, high vibrations, high power consumption, poor surface finish, affecting surface integrity, etc. The topics included in the conference will provide a platform to researchers, practicing professionals from industry to present and deliberate upon their research work on various topics to understand the current status of the same in India. There is a need to create awareness among the professionals in this field of engineering about the need to adopt sustainability principles into the machining processes and contribute to the progress in this field, reduce pollution, conserve energy and reduce its environmental effects.

I place on record my strong appreciation and thanks to the All India Council for Technical Education (AICTE), New Delhi, for sponsoring this two-day e-Conference. I also thank NMAMIT, Nitte, for providing other support for organizing this e-Conference in a befitting manner. In spite of the conference theme being very specialized and the prevailing pandemic situation, the response in terms of number of papers received has been very good. I thank all the authors of the papers, their organizations and participants for supporting this e-Conference. **I thank M/s Springer Nature for agreeing to publish all the accepted papers of this conference in the *Lecture Notes in Mechanical Engineering* (LNME)**. I thank the keynote speaker, Dr. P. V. Rao, from IIT Delhi and all the invited speakers, who readily agreed to our request and for presenting expert talks on themes of relevance to this conference. I will be failing in my duty if I do not thank all the paper reviewers, who responded to our request on short notice and helped us in the review process. I thank our dear Principal Dr. Niranjan N. Chiplunkar for supporting and guiding us in the conduct of this conference. I thank our dear HOD, Dr. Shashikantha Karinka, for helping us in all possible ways. I thank the Vice Principals, Deans and other officials for guiding us. A special thanks to Dr. Grynal D'mello and Mr. Bhaskara P. Achar, the conveners of this conference, who stood as strong pillars behind me and supported me in this effort. I thank all the organizing committee members and all others who have helped us directly or indirectly in organizing this conference. I hope and wish this conference will benefit all the stakeholders and help in generation and dissemination of knowledge related to the theme, which is an important area of interest to the manufacturing industries, which use machining operations.

Nitte, India

Dr. P. Srinivasa Pai
Organizing Secretary
National e-Conference on Sustainable Machining
Strategies for Better Performance (NCSMSBP 2020)

# Contents

**Taguchi Experimental Design for Turning of AISI 4340 Steel and Grey Analysis on Machinability Parameters for Sustainable Machining** . . . . . 1
Gautam S. Shetty and Gajanan M. Naik

**Development of 3D Printed Electromyography Controlled Bionic Arm** . . . . . 11
Shiv Pratap Singh Yadav, Vijay Kumar Shankar, L. Avinash, Abdulrajak Buradi, B. A. Praveena, Vikram Kedambi Vasu, N. Vinayaka, and K. Dilip Kumar

**Multi-response Optimization of Machining Characteristics Using MQL Through GRA and TOPSIS Approach** . . . . . 23
A. Venkata Vishnu, S. Sudhakar Babu, and P. Jamaleswara Kumar

**An Experimental Investigation of Laser-Assisted Machining of EN24 Steel** . . . . . 39
Ajit M. Hebbale, S. Rajesh K. Reddy, Mirza Abdul Hadi Baig, Manish Tak, and Ravi N. Bathe

**Optimization of Wire Cut Electric Discharge Machining Characteristics of Hybrid Aluminium Composites (Al6061/Gr/SiCp) Using Taguchi Method** . . . . . 49
P. Gavisiddesha, C. Thotappa, Veerabhadrappa Algur, and B. Suresh Reddy

**Minimum Quantity Lubrication and Cryogenic for Burnishing of Difficult to Cut Material as a Sustainable Alternative** . . . . . 61
B. Sachin, Charitha M. Rao, Gajanan M. Naik, C. Durga Prasad, Ajit M. Hebbale, V. Vijeesh, and Muralidhara Rao

**Investigation of Effect of EDM Process Variables on Material Removal Rate and Tool Wear Rate in Machining of EN19 Steel Using Response Surface Methodology** ... 71
Santosh Nandurkar, Sachin Kulkarni, Tushar Hawal, Niranjan Pattar, and Nagaraj Kelageri

**Sustainability Analysis of Cutting Fluids in Minimum Quantity Lubrication of Machining Operations** ... 83
P. Jamaleswara Kumar and B. V. S. Arun Kumar

**The Effect of Drilling Parameters on the Hole Quality of Hybrid Fiber-Reinforced Epoxy Composite** ... 99
V. Santhanam, S. Sendhilkumar, N. Venkateshwaran, and M. Chandrasekaran

**Effect of Profile Geometry and Cutting Speed Override Parameter on Profiling Speed During Tapering Using Wire Electric Discharge Machining** ... 111
I. V. Manoj and S. Narendranath

**Evaluation of Machining Properties of Short Bamboo Fiber-Based Green Composites Using CNC Drilling Process** ... 123
Shubham B. Patil, Jagadish, Shailesh Vaidya, and Satish Kumar Adapa

**Influence of Burnishing Process on Tensile Strength of Al7075-T6 Alloy** ... 133
Pavana Kumara and Udaya Prasanna Handadi

**Comparison of GRA and TOPSIS Optimization Techniques in DMLS-Processed Bronze–Nickel Samples** ... 143
R. Rajesh, Mithun V. Kulkarni, P. Sampathkumaran, P. Sathish, and S. Sreenivas

**Optimization of Parameters for Material Removal Rate and Surface Roughness in Wire Electric Discharge Grinding (WEDG) for Micro-machining of Cemented Carbide Rods** ... 161
M. Parthiban and M. Harinath

**Exploration of Effectiveness of Ionic Liquid Adopted as an Additive to the Vegetable Oils** ... 171
Harpreet Singh, Balraj Singh, and Roshan Lal Virdi

**Experimental Investigation of Vegetable Oils-Based Minimum Quantity Lubrication Grinding by Using Ionic Liquid** ... 185
Balraj Singh, Harpreet Singh, Roshan Lal Virdi, and Khushdeep Goyal

**Comparison of Copper and Tungsten Electrodes for the Electric Discharge Machined SUS-316L** ... 197
Gurpreet Singh, Amit Mahajan, Sandeep Devgan, and Sarabjeet Singh Sidhu

**Surface Integrity of Powder Mixed Electrical Discharge Treated Substrate at High Discharge Energies** ..... 207
Sandeep Devgan, Amit Mahajan, Gurpreet Singh, Gurcharan Singh, and Sarabjeet Singh Sidhu

**Analysis of Effect of Machining Parameters on Surface Roughness and MRR of AA3003/SiC Composite Material** ..... 219
Sachinkumar Patil, M. Nagamadhu, K. Anand Babu, S. B. Kivade, and T. Veerbhadrappa

**Use of Vortex Tube Cooling for Machining Stellite 6** ..... 227
G. Benaka, Bhaskara P. Achar, P. Srinivasa Pai, Grynal D'mello, and K. G. Gururaj

# About the Editors

**Dr. P. Srinivasa Pai** is currently working as Professor in the Department of Mechanical Engineering at NMAMIT, Nitte, Karnataka. He obtained his B.E. in Industrial and Production Engineering, from NMAMIT, Nitte in 1993 and M.Tech. in Maintenance Engineering, from SJCE, Mysore in 1996. He obtained his Ph.D in Mechanical Engineering, from Mysore University in 2004. He went to Utah State University, Utah, USA, as a Post-Doctoral Research Scholar in Metal Machining for a period of 1 year during 2009–2010. Some of his areas of interests are high speed machining, surface integrity studies, difficult to machine materials, condition monitoring, artificial neural networks, signal processing, optimization techniques and engineering education. He has 200 papers in various National and International Journals and Conference proceedings. He has won the following awards: Indian Society for Technical Education (ISTE)—SGSITS National Award: for The Research Work done by Young Teachers (Below 35) of Engineering Colleges for the year 2005 and Air India BOLT (Broad Outlook—Learner Teacher) Award for the District of Udupi for the year 2006. He won the Appreciation Certificate from ISTE, New Delhi for guiding the II Best M.Tech. Thesis in Mechanical Engineering during 2012. He won the "Best Researcher Award", instituted by NMAMIT, Nitte in 2013. Currently he is in the Editorial Advisory Board of *International Journal of COMADEM* and Section Editor (Mechanical and Materials Engineering) of *Indian Journal of Engineering*. He has guided seven Ph.D. scholars, one is going to submit shortly. Currently he is guiding one Ph.D. scholar.

**Dr. V. Krishnaraj** is currently working as Professor in Production Engineering and Head of Advanced Tool & Die Center, PSG College of Technology, Coimbatore, Tamil Nadu. He obtained his B.E. in Mechanical Engineering, from CIT, Coimbatore in 1994 and M.E. in Production Engineering, from PSG College of Technology, Coimbatore in 1999. He obtained his Ph.D. in Machining of Composites, from Anna University, Chennai in 2007. He did his Post-Doctoral Research Fellowship in Machining for a period of one year during 2008–2009 from University Paul Sabatier, France and was also a research fellow in UNSW Australia. His areas of research interests are composite materials, metal cutting and tool design. He has more than 90 publications in international journals. He has several books and book chapters to his credit and has guided nine Ph.D. theses.

# Taguchi Experimental Design for Turning of AISI 4340 Steel and Grey Analysis on Machinability Parameters for Sustainable Machining

**Gautam S. Shetty and Gajanan M. Naik**

**Abstract** In presentt-day scenario, energy-efficient product or process is highly in demand. In the present work, we have made an attempt to study the effects of machining parameters in turning process of AISI 4340 steel by using grey relational analysis. Experiments were performed using Taguchi experimental design to reduce the number of experiments in the study. Taguchi's L9 orthogonal array was chosen for design of experiments. Experiments were done by carrying out turning operations on dry cutting conditions by using high-quality multilayer-coated titanium carbide tool insert. Cutting speed, feed and depth of cut were chosen as the machinability parameters. The aim of grey analysis is to get the best set of parameters for getting optimized value of material removal rate (MRR) and surface roughness. The levels of cutting speed chosen were 160, 200 and 240 m/min, feed levels were 0.14, 0.18 and 0.22 mm/rev, and depths of cut were 0.5, 0.75 and 1 mm, respectively. The results of grey analysis showed that cutting speed 240 m/min, feed 0.5 mm/rev and depth of cut 0.75 gave the optimized value for MRR and surface roughness. Hence, using both grey analysis and Taguchi technique, we can considerably reduce the energy consumption by improving the machining process making the process easier and smoother to machine and energy efficient by eliminating the unnecessary effort.

**Keywords** Taguchi L9 array · AISI 4340 · Grey analysis · Design of experiments · Surface roughness · Material removal rate (MRR)

## 1 Introduction

Energy-efficient process and techniques are in demand in modern day as energy consumption is going very high due to the growing global population. Hence, it has become very important necessities for industries to reduce their energy consumption for saving capital as the cost for energy is skyrocketing. Steel and its products one of the widespread uses in domestic transportation construction and other utility

G. S. Shetty (✉) · G. M. Naik
Mechanical Department, Mangalore Institute of Technology and Engineering, Mangalore, Karnataka 574225, India

 
P. Srinivasa Pai and V. Krishnaraj (eds.), *Sustainable Machining Strategies for Better Performance*, Lecture Notes in Mechanical Engineering,
https://doi.org/10.1007/978-981-16-2278-6_1

products. Industries try to expand focus on increasing the productivity and quality of products. Quality also plays an important role in the performance of the products produced. Machining makes one of the major finishing processes for most of the products of steel. Quality and productivity in machining is measured by surface roughness (Ra) and material removal rate (MRR) [1–3]. Davis et al. [2] performed experiments to analyse cutting parameters for EN24 steel, and an attempt was made to optimize the cutting parameters. In this, the turning operation was carried out on EN24 steel by carbide P-30 cutting tool in dry condition, and the combination of the optimal levels of the parameters was obtained. Sahoo et al. [4] applied response surface methodology on investigating flank wear in machining hardened steel using PVD TiN-coated mixed ceramic insert. The experiments have been conducted using three-level full factorial design techniques. The machinability model has been developed in terms of cutting speed ($v$), feed ($f$) and machining time ($t$) as input variable using response surface methodology. The adequacy of model has been checked using correlation coefficients. As the determination coefficient, R2 (98%) is higher for the model developed; the better is the response model fits the actual data. In addition, residuals of the normal probability plot lie reasonably close to a straight line showing that the terms mentioned in the model are statistically significant. The predicted flank wear has been found to lie close to the experimental value. This indicates that the developed model can be effectively used to predict the flank wear in the hard turning. Abrasion and diffusion have been found to be the dominant wear mechanism in machining hardened steel from SEM micrographs at highest parametric range. Machining time has been found to be the most significant parameter on flank wear followed by cutting speed and feed as observed from main effect plot and ANOVA study; this is because of the long contact duration between tool and work piece. Tzeng et al. [5] have studied the optimization of CNC turning process parameters for SKD11 by GRA and found that factor depth of cut is the most significant parameter of the turning operation. Maiyar et al. [6] reported the multi-response conditions based on the Taguchi OA with GRA. This study helps us to understand the technique. Several studies on Taguchi optimization and grey relational analysis were carried out by Shetty et al. [7], Vishwas et al. [8, 9], Naik et al. [10], Sachin et al. [11]. Current study investigates the multi-response optimization of CNC machining parameters through Taguchi–grey relational analysis for AISI 4340 steel by considering cutting speed, feed and depth of cut as input parameters and material removal rate and surface roughness as output responses. Further, the effect of input parameters on output responses has been studied through grey relational grade mean plot.

## 2 Experimental Details

AISI 4340 steel with a dimension of 100 mm length and 30 mm diameter has been taken for machining and machined under dry condition with turning length of 10 mm. The machining was carried out according to Taguchi L9 array experiments using CNC lathe. Kennametal make: Titanium-Coated Carbide Inserts CNMG 120,408

RP KC9110 shown in Fig. 1b was used as cutting tool used in this present project work. The dimensions of specimens for spectrometer analysis were prepared to size of 31 mm diameter and height 10 mm. The composition analysis of the AISI 4340 is presented in Table 1. In this present study, we have selected three levels and three factors. The number of levels selected for each factor is three. Table 2 presents the machining parameter with their levels. The cutting speed, feed and depth of cut are the three parameters (factors) selected in the experiment.

The turning operation were carried out in CNC machine of Fanuc make with the specification's maximum turning diameter: 100 mm, spindle speed: 60–6000 rpm, spindle power: 5.5/3.7 KW, *X*, *Y*—travel: 250 mm. Table 3 shows the Taguchi L9 array for the machining parameters for AISI 4340 steel with a set of experiments that has to be conducted with corresponding values. Minitab statistical software package is used to design the experiments. Surface roughness was measured using Talysurf (SJ-301) with evaluation length 2.4 mm, and material removal rate (MRR)

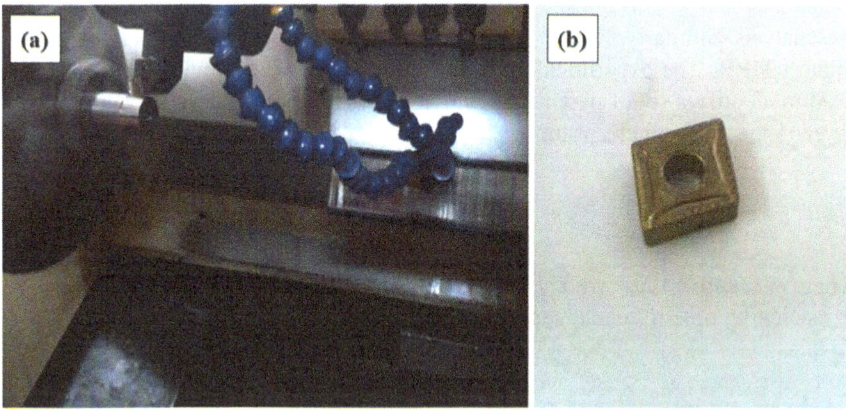


**Fig. 1** **a** Turning and facing of specimen in CNC lathe in dry conditions and **b** single-point cutting tool insert

**Table 1** Composition analysis of AISI 4340

| Elements | C % | Fe % | Mo % | Ni % | Cr % | Si % | Mn % | P % | S % |
|---|---|---|---|---|---|---|---|---|---|
| AISI 4340 | 0.331 | 95.6 | 1.59 | 1.28 | 0.93 | 0.193 | 0.52 | 0.033 | 0.026 |

**Table 2** Machining parameters and its levels

| Machining parameters | Symbols | Levels | | |
|---|---|---|---|---|
| | | 1 | 2 | 3 |
| Cutting speed (m/min) | A | 160 | 0.14 | 0.5 |
| Feed rate (mm/rev) | B | 200 | 0.18 | 0.75 |
| Depth of cut (mm) | C | 240 | 0.22 | 1.0 |

**Table 3** Taguchi L9 array for machining parameters

| Exp. No. | Cutting speed (m/min) | Feed rate (mm/rev) | Depth of cut (mm) |
|---|---|---|---|
| 1 | 160 | 0.14 | 0.5 |
| 2 | 160 | 0.18 | 0.75 |
| 3 | 160 | 0.22 | 1 |
| 4 | 200 | 0.14 | 0.75 |
| 5 | 200 | 0.18 | 1 |
| 6 | 200 | 0.22 | 0.5 |
| 7 | 240 | 0.14 | 1 |
| 8 | 240 | 0.18 | 0.5 |
| 9 | 240 | 0.22 | 0.75 |

was measured using Eq. (1). In our analysis, surface roughness (Ra) a condition of 'smaller is better' was chosen since our goal is to achieve the lowest Ra value in the analysis. Similarly, 'larger is better' is selected for MRR as we want to get the highest MRR. The experimental results were studied with grey analysis, which is used to identify a set of parameters which will give optimized set of values for the given range of parameter values.

$$\text{MRR} = \frac{\text{Wt}_\text{b} - \text{Wt}_\text{a}}{t * \rho} \tag{1}$$

where $\text{Wt}_\text{b}$ and $\text{Wt}_\text{a}$ are weight before and after machining in grams, respectively. $t$ is machining time in second, and $\rho$ is density in grams/cc.

## 3 Results and Discussion

In the present work, a total of 27 experiments were done; i.e.: Taguchi L9 array has nine experiments but we need to check for consistency the experiments were further more repeated for 2 more trails as shown in Table 3. Hence, experiments were conducted in three trials/runs. These 27 experiments were performed in CNC machine. Turning operation was carried out on each individual specimen to a length of 10 mm. For each experiment, three values for surface roughness were measured on the specimen. Table 4 gives the average value of Ra and MRR in thee trials of one experiment.

It is known easily that higher surface finish is obtained for lower value of surface roughness and from Table 4 that lower value of surface roughness is observed for cutting speed, feed and depth of cut 200 m/min, 0.14 mm/rev and 0.75 mm, respectively, with an average value of 0.668 μm. Also, it is observed that the highest surface roughness is found for cutting speed, feed and depth of cut 160 m/min, 0.22 mm/rev and 1 mm with average value of 1.676 μm, respectively.

**Table 4** Surface roughness and MRR averages in three trial runs

| Exp. No. | Cutting speed (m/min) | Feed (mm/rev) | Depth of cut (mm) | $Ra_1$ (μm) | $Ra_2$ (μm) | $Ra_3$ (μm) | $Ra_{avg}$ (μm) | MRR 1 ($mm^3$/sec) | MRR 2 ($mm^3$/sec) | MRR 3 ($mm^3$/sec) | MRR ($mm^3$/sec) |
|---|---|---|---|---|---|---|---|---|---|---|---|
| 1 | 160 | 0.14 | 0.5 | 1.263 | 1.187 | 1.210 | **1.220** | 174.51 | 223.37 | 242.04 | **213.31** |
| 2 | 160 | 0.18 | 0.75 | 1.383 | 1.287 | 1.283 | **1.318** | 372.76 | 361.74 | 361.74 | **365.41** |
| 3 | 160 | 0.22 | 1 | 1.723 | 1.667 | 1.637 | **1.676** | 525.94 | 532.71 | 615.81 | **558.16** |
| 4 | 200 | 0.14 | 0.75 | 0.687 | 0.667 | 0.650 | **0.668** | 355.43 | 377.37 | 377.37 | **370.05** |
| 5 | 200 | 0.18 | 1 | 1.253 | 1.207 | 1.167 | **1.209** | 597.15 | 559.83 | 597.15 | **584.71** |
| 6 | 200 | 0.22 | 0.5 | 1.517 | 1.487 | 1.490 | **1.498** | 665.46 | 406.95 | 443.64 | **505.35** |
| 7 | 240 | 0.14 | 1 | 0.910 | 0.957 | 0.883 | **0.917** | 570.01 | 492.50 | 562.26 | **541.59** |
| 8 | 240 | 0.18 | 0.5 | 1.350 | 1.147 | 1.180 | **1.226** | 496.38 | 443.64 | 465.82 | **468.61** |
| 9 | 240 | 0.22 | 0.75 | 1.573 | 1.490 | 1.473 | **1.512** | 653.13 | 683.51 | 683.51 | **673.39** |

For better MRR, we need to choose higher value. It is also observed from Table 4 that the best values for MRR are observed for cutting speed, feed and depth of cut 240 m/min, 0.22 mm/rev and 0.75 mm with an average value of 673.39 $mm^3$/sec. Similarly, lowest MRR is observed for 160 m/min, 0.14 mm/rev and 0.5 mm speed, feed and depth of cut with average value of 213.31 $mm^3$/s, respectively. While optimizing through Taguchi's analysis, single response can be optimized, so that, achieved result contributed better MRR and low surface roughness but in grey relational analysis, instead of having two optimized setting, individually for each response grey providing single optimum parameter results for getting good MRR and surface roughness. Hence, to achieve the above result, the following steps have been followed.

With the help of Taguchi L9 array, we were able to reduce the number of experiments in the work. In order to do grey analysis on the results of the experiments, we need to get the values for normalized value for the response parameters. Table 5 gives the minimum and maximum values for these parameters. By standard formula to get the normalized value of response parameters, we use the [8, 9].

$$X_{ij} = \frac{y_{ij} - \min_j y_{ij}}{\max_j y_{ij} - \min_j y_{ij}} \tag{2}$$

$$X_{ij} = \frac{\max_j y_{ij} - y_{ij}}{\max_j y_{ij} - \min_j y_{ij}} \tag{3}$$

where $x_{ij}$ is normalized results and $y_{ij}$ is the $i$th experimental results in the $j$th experiment. The above formula (2) for MRR and Eq. (3) for Ra was used, and normalized Ra and MRR values are found, respectively, and are presented in Table 5. Minimum and maximum deviation sequence values are given in Table 5. The deviation sequence or delta is calculated by subtracting normalized values from 1 (unity). Grey relation coefficient is calculated by using the formula (4) as given below [8, 9].

$$\text{Grey relational coefficient } \gamma(j) = \frac{\Delta_{\min} + \xi\Delta_{\max}}{\Delta_{i(r,i)} + \xi\Delta_{\max}} \tag{4}$$

where $\Delta(r,i)$ is the deviation sequence values of $i$ trial, $\Delta_{\min}$ is the smallest value of the deviation sequence of $\Delta(r,i)$ for $j$th response, $\Delta_{\max}$ is the largest value of $\Delta$ $(r, i)$ for $j$th response, $\xi$ is distinguishing coefficient in the current work we have set this value to 0.5 since we have given equal weightage to Ra and MRR. The grey relational grade value is determined by taking average values of Ra and MRR grey relational coefficients (GRC). All the values calculated are presented in Table 6. Table 6 shows the average value for GRG for cutting speed, feed and depth of cut for their corresponding levels, respectively. The maximum and minimum average values were noted, and the difference between them gave the optimum level for each input parameter for the desired optimized value for Ra and MRR. Hence, with a value of cutting speed, feed and depth of cut 240 m/min, 0.14 mm/rev and 0.75 mm would give the optimized value for the Ra and MRR, respectively. Figure 2 shows

**Table 5** Surface roughness and MRR grey analysis

| Exp. No. | $Ra_{avg}$ (μm) | MRR ($mm^3$/sec) | Normalized Ra | Normalized MRR | Deviation sequence (Delta) | | Grey relational coefficient (GRC) | | GRG (Avg) | Rank |
|---|---|---|---|---|---|---|---|---|---|---|
| | | | | | Ra | MRR | Ra | MRR | | |
| 1 | 1.220 | 213.308 | 0.4520 | 0.0000 | 0.5480 | 1.0000 | 0.4771 | 0.3333 | 0.4052 | 9 |
| 2 | 1.318 | 365.412 | 0.3550 | 0.3306 | 0.6450 | 0.6694 | 0.4367 | 0.4276 | 0.4321 | 8 |
| 3 | 1.676 | 558.156 | 0.0000 | 0.7495 | 1.0000 | 0.2505 | 0.3333 | 0.6663 | 0.4998 | 6 |
| **4** | **0.668** | **370.053** | **1.0000** | **0.3407** | **0.0000** | **0.6593** | **1.0000** | **0.4313** | **0.7156** | **1** |
| 5 | 1.209 | 584.710 | 0.4631 | 0.8073 | 0.5369 | 0.1927 | 0.4822 | 0.7218 | 0.6020 | 4 |
| 6 | 1.498 | 505.349 | 0.1764 | 0.6348 | 0.8236 | 0.3652 | 0.3778 | 0.5779 | 0.4778 | 7 |
| 7 | 0.917 | 541.590 | 0.7530 | 0.7135 | 0.2470 | 0.2865 | 0.6694 | 0.6358 | 0.6526 | 3 |
| 8 | 1.226 | 468.613 | 0.4465 | 0.5549 | 0.5535 | 0.4451 | 0.4746 | 0.5291 | 0.5018 | 5 |
| 9 | 1.512 | 673.386 | 0.1621 | 1.0000 | 0.8379 | 0.0000 | 0.3737 | 1.0000 | 0.6869 | 2 |
| Min | 0.668 | 213.308 | | | 0 | 0 | | | | |
| Max | 1.676 | 673.386 | | | 1 | 1 | | | | |

**Table 6** Response table for GRG

| Parameters | Level 1 | Level 2 | Level 3 | Max | Min | Diff | Optimum level |
|---|---|---|---|---|---|---|---|
| Cutting speed (m/min) | 0.4457 | 0.5985 | **0.6138** | 0.6138 | 0.4457 | 0.1680 | **A3** |
| Feed (mm/rev) | **0.5911** | 0.5120 | 0.5548 | 0.5911 | 0.5120 | 0.0792 | **B1** |
| Depth of cut (mm) | 0.4616 | **0.6115** | 0.5848 | 0.6115 | 0.4616 | 0.1499 | **C2** |

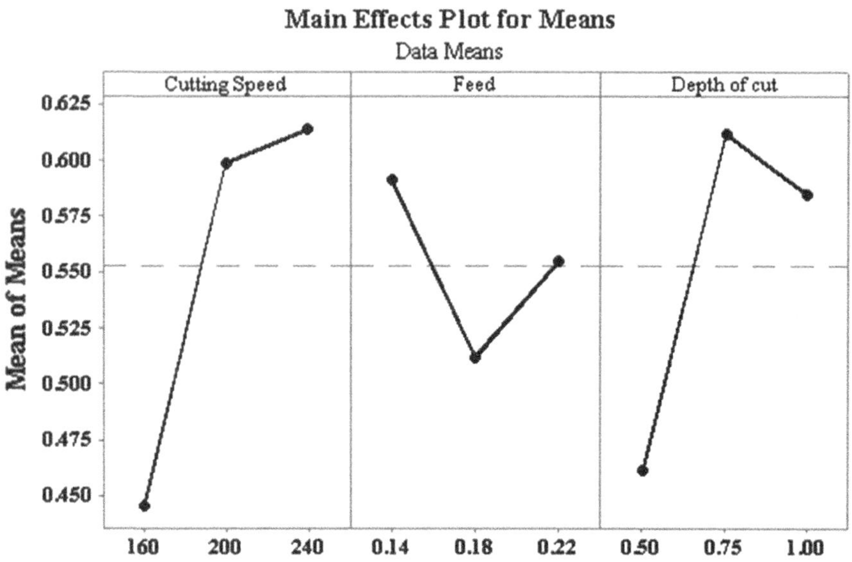


**Fig. 2** Main effect plot for GRG

the main effects plot for GRG versus the input parameters, also agrees with the levels we mentioned above. The GRG values for the above parameters levels are highest in the graph.

Chip morphology has been done to identify the nature of chip–tool interaction influenced by using coated carbide inserts. Chips were collected after every experiment run, and its shapes and colours were examined by digital camera to identify the nature of chip–tool interaction. The chips found are shown in Fig. 3a is for experiment no. 4 which has lowest surface roughness value is found to have continuous chips, whereas Fig. 3b shows the chips for experiment no. 9 which has highest MRR has discontinuous chips.

Further, an optimum parameter of getting good material removal rate and low surface roughness was estimated through Taguchi-based grey relational analysis. However, the obtained optimal parameters were not present within the L9 orthogonal array of experimental trials. Hence, the confirmation test has been conducted by considering optimal level of process parameters. Table 7 presents the experimental

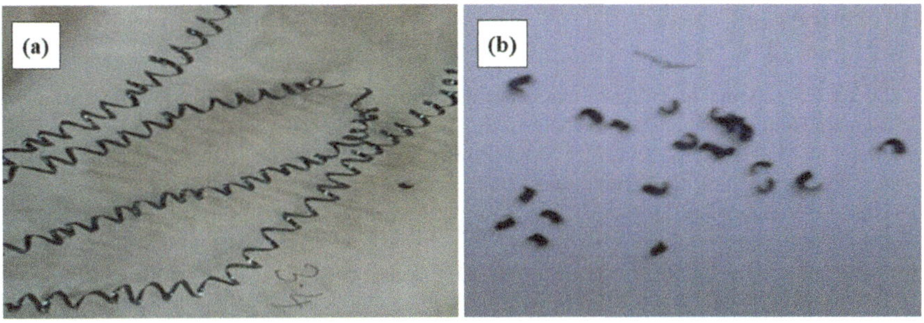


**Fig. 3** **a** Chips for highest surface roughness (#4) and **b** chips for highest MRR (#9)

**Table 7** Confirmation test results

| Condition | Taguchi–grey theory prediction (4th trial of OA) | Experimental | Percentage change (%) |
|---|---|---|---|
| Optimum level | A2 B1 C2 | A3 B1 C2 | – |
| Ra (μm) | 0.668 | 0.522 | −24.86% |
| MRR ($mm^3$/s) | 370.053 | 439.320 | 15.77% |

test results of 4th trial of orthogonal array which is ranked first in Table 5 and GRA optimum setting parameters after conducting experiments for material removal rate and surface roughness of CNC turning for AISI 4340 steel. It can be seen from the obtained results that the percentage increase of material removal rate and percentage decrease of surface roughness are about 15.77% and 24.86%, respectively.

## 4 Conclusions

In the current project he had attempted to optimize the surface roughness and MRR value for given range of input parameters it is found that experiment no. 4 gave the best desired result with rank 1, also experiment no. 1 gave the least desired result and rest of the rank are presented in the Table 5 with cutting speed, feed and depth of cut 200 m/min, 0.14 mm/rev and 0.75 mm, respectively, has the most optimized result in the designed set of experiments. Further, the response table for grey relational grade (GRG) shows that the best possible optimization would be obtained for input parameters of A3, B1 and C2, i.e. cutting speed, feed and depth of cut 240 m/min, 0.14 mm/rev and 0.75 mm, respectively. Hence, it is understood that the highest speed, lowest feed and moderate depth of cut levels of parameters provide best desired output of optimized Ra and MRR. It is clearly observed that better surface finish has longer helical continuous chips, whereas better MRR has loose small continuous chips. Further investigation can be carried out for wet machining conditions by using cryogenic fluids for new range of input parameters.

## References

1. Shetty GS (2020) The effect of machinability parameters on AISI 4340 (EN-24) steel using Taguchi technique. Int J Mech Prod Eng Res Dev 10:302–310
2. Davis R, Madhukar JS, Rana VS, Singh P (2012) Optimization of cutting parameters in dry turning operation of EN24 steel. Int J Emerg Technol Adv Eng 2(10):1–10
3. Naik GM, Narendranath S (2017) A parametric optimization of wire-ED turning process parameters on material removal rate of INCONEL 718. J Mech Eng Biomech 2(2):8–14
4. Sahoo AK, Sahoo B (2011) Surface roughness model and parametric optimization in finish turning using coated carbide insert: response surface methodology and Taguchi approach. Int J Ind Eng Comput 5:819–830
5. Tzeng C-J, Lin Y-H, Yang Y-K, Jeng M-C (2009) Optimization of turning operations with multiple performance characteristics using the Taguchi method and Grey relational analysis. J Mater Process Technol 209(6):2753–2759
6. Maiyar LM, Ramanujam R, Venkatesan K, Jerald J (2013) Optimization of machining parameters for end milling of Inconel 718 super alloy using Taguchi based grey relational analysis. Procedia Eng 64:1276–1282
7. Shetty AS, Akshar KS, Prashanth BY, Naik GM (2017) Optimization of machining parameters on MRR for EN19 & EN31 steel using Taguchi method. Int J Emerg Res Manage 6:2278–9359
8. Vishwas CJ, Girish LV, Naik GM, Sachin B, Roy A, Prashanth BY, Badiger R (2018) Effect of machining parameters on surface integrity during dry turning of AISI 410 martensitic stainless steel. IOP Conf. Ser.: Mater. Sci. Eng. 376(1):012127
9. Vishwas CJ, Gajanan MN, Sachin B, Abhinaba R, Puneet NP, Anjan BN, Vinayak NK (2019) Study on surface roughness in minimum quantity lubrication turning of Al-6082/SiC metal matrix composites. Appl Mech Mater 895:127–133
10. Naik GM, Narendranath S (2018) Optimization of wire-ED turning process parameters by Taguchi-Grey relational analysis. i-Manager's J Mech Eng 8(2):1–10
11. Sachin B, Narendranath S, Chakradhar D (2018) Experimental evaluation of diamond burnishing for sustainable manufacturing. Mater Res Express 5(10):106514

# Development of 3D Printed Electromyography Controlled Bionic Arm

**Shiv Pratap Singh Yadav, Vijay Kumar Shankar, L. Avinash, Abdulrajak Buradi, B. A. Praveena, Vikram Kedambi Vasu, N. Vinayaka, and K. Dilip Kumar**

**Abstract** Humanoid autonomy is a creating zone of exploration on account of its likely applications in orthotics and prosthesis for people. With the correct now open advancements, the most uncommon, mechanized hands used in prosthetics can cost a huge amount, making it closed off to everyone of amputees and apply self-rule experts. An enormous segment of the features gave by these expensive advancements are trivial to various customers, making a phenomenal gap in cost and organizations among customers and development. Using the rising 3D printing advancement, our endeavor is to manufacture a 3D printed robotized hand that can recreate a similar number of fundamental functionalities of the advanced exorbitant hands, while restricting the cost. Convinced by the noteworthy cost of bleeding edge humanoid computerized hand, the target of this undertaking expected is to assemble a sensible 3D printed mechanical hand that can copy a similar number of the functionalities of the pushed hand, while restricting the cost. Finally, the developed 3D printed electromyography controlled bionic arm gives the right choice to enhance the observation of the mechanism.

**Keywords** EMG sensor · Electromyography · Additive manufacturing · Fused deposition modeling · Servo motor · Acrylonitrile butadiene styrene (ABS)

S. P. S. Yadav (✉) · V. K. Shankar · L. Avinash · A. Buradi · B. A. Praveena · V. K. Vasu
Department of Mechanical Engineering, Nitte Meenakshi Institute of Technology, Bengaluru, Karnataka 560064, India

N. Vinayaka
Department of Aeronautical Engineering, Nitte Meenakshi Institute of Technology, Bengaluru, Karnataka 560064, India

K. D. Kumar
Department of Mechanical Engineering, NMAM Institute of Technology Nitte, Karkala Taluk, Udupi, Karnataka 574110, India



P. Srinivasa Pai and V. Krishnaraj (eds.), *Sustainable Machining Strategies for Better Performance*, Lecture Notes in Mechanical Engineering,
https://doi.org/10.1007/978-981-16-2278-6_2

## 1 Introduction

Humanoid mechanical technology is a developing territory of research because of its potential applications in orthotics and prosthesis for people [1]. The human hand is comprised of 27 bones, 34 muscles, and 123 tendons; the human hand is an unpredictable framework that suggests a difficult research conversation starter to humanoid apply autonomy on the most proficient method to reproduce it with mechanical parts body [2, 3]. For quite a long time, trailblazer has been attempting to supplant lost appendages with artificial gadgets. A few prosthetic gadgets have been found from antiquated civilizations around the globe exhibiting the continuous advancement of prosthetic innovation. Humanoid mechanical autonomy can perform a large portion of the errands that a human hand can perform, going from getting objects of different shapes and weight dissemination, controlling with instruments to perform complex undertakings, getting little, and delicate article [4].

After some time, materials improved, and plans began joining pivots and pulley frameworks. This prompted basic mechanical body-controlled gadgets, for example, metal snares which can open and close as a client twists their elbow for instance. The human hand contains in any event 27 bones 34 muscles which move the fingers and thumb and numerous tendons, nerves, and supply routes. Prostheses expect to imitate the elements of the human body and return usefulness to people with missing furthest points [5]. The human hand has 27 degrees of opportunity: 4 in each finger, 3 for expansion and flexion [6], and one for kidnapping and adduction; the thumb is increasingly confused and has 5 DOF, leaving 6 DOF for the pivot, and interpretation of the wrist [7].

Electric prostheses, likewise, generally alluded to as myoelectric prostheses, are controlled utilizing electric signals that are really made by your body's muscles [8]. These prosthetics work by utilizing your current muscles in your lingering appendage to control the elements of the prosthetic gadget itself [9]. Body-controlled prostheses work by utilizing links to interface the development of the body to the prosthesis and to control it. Moving the body with a specific goal in mind will pull on the link and cause it to open, close, or curve. A three-saddle link framework prosthesis permits the wearer to get a handle on items, flex, and lock the elbow [10]. This is the most affordable kind of arm prosthesis.

## 2 Methodology

A kinematic ideal has stayed produced for replication of the prehensile capacities of the mortal hand. The kinematic frame of the hand is described by seams and basic sections as shown in Fig. 1. The prototypical depends on a calculation that decides interaction between two ellipsoids, which are developed to estimate the geometry of the cutaneous superficial of the hand slices. The conventional forecast the hand posture for power grasp of ellipsoid item by enfolding the limbs around

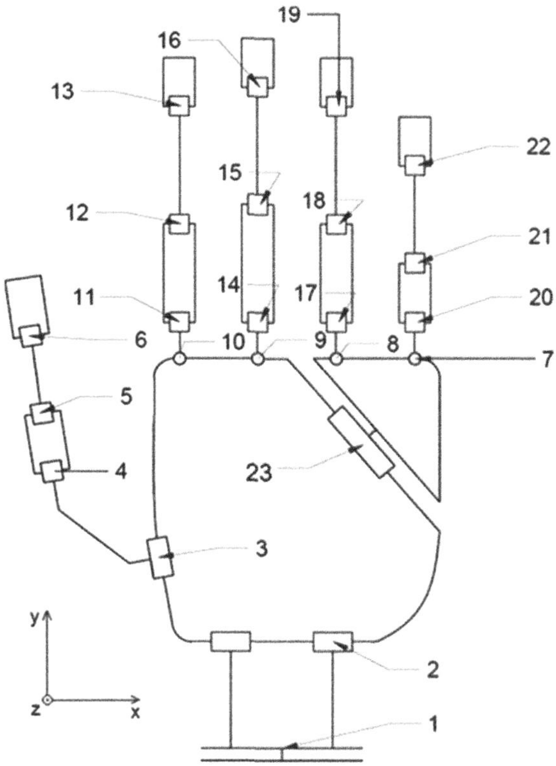


**Fig. 1** Kinematic frame of hand

the item. Coefficients for assessing anthropometric boundaries from pointer extent and broadness are consolidated in the prototypical [11]. Designs techniques are incorporated for graphical presentation of the archetypal. With an end goal to approve the prescient capacities of the typical, combined edges remained estimated on six themes getting a handle on round chambers of different diameters across and these deliberate joint angles were contrasted and points anticipated by the model [12].

Except if unequivocally expressed something else, automated systems are frameworks of inflexible bodies associated by joints. The position and direction of an inflexible body is spaces are all things considered named the "present." In this way, robot kinematics portrays the posture, speed, acceleration, and all higher-order derivatives of the posture of the bodies that contain a system [13]. Table 1 lists number of joints with angular rotation of each joint.

The synchronize systems of mortal pointer are shown. For final undistinguishable transformation matrix as shown by Eqs. (1)–(3), following matrix is used as basic matrices for turning motion [14]:

**Table 1** Rotation of particular joint

| Number of joints | Angle (°) | Number of joints | Angle (°) |
|---|---|---|---|
| 1 | 120 | 12 | 100 |
| 2 | 120 | 13 | 80 |
| 3 | 30 | 14 | 90 |
| 4 | 35 | 15 | 100 |
| 5 | 90 | 16 | 80 |
| 6 | 90 | 17 | 90 |
| 7 | 30 | 18 | 100 |
| 8 | 30 | 19 | 80 |
| 9 | 30 | 20 | 90 |
| 10 | 30 | 21 | 100 |
| 11 | 90 | 22 | 80 |

$$R_x = \begin{bmatrix} 1 & 0 & 0 & 0 \\ 0 & \cos\phi & -\sin\phi & 0 \\ 0 & \sin\phi & \cos\phi & 0 \\ 0 & 0 & 0 & 1 \end{bmatrix} \tag{1}$$

$$R_y = \begin{bmatrix} \cos\varphi & 0 & \sin\varphi & 0 \\ 0 & 1 & 0 & 0 \\ \sin\varphi & 0 & \cos\varphi & 0 \\ 0 & 0 & 0 & 1 \end{bmatrix} \tag{2}$$

$$R_z = \begin{bmatrix} \cos\vartheta & -\sin\vartheta & 0 & 0 \\ \sin\vartheta & \cos\vartheta & 0 & 0 \\ 0 & 0 & 1 & 0 \\ 0 & 0 & 0 & 1 \end{bmatrix} \tag{3}$$

For linear motion,

$$T_0^1 = \begin{bmatrix} 1 & 0 & 0 & I_x \\ 0 & 1 & 0 & I_y \\ 0 & 0 & 1 & I_z \\ 0 & 0 & 0 & 1 \end{bmatrix} \tag{4}$$

By utilizing of fundamental referenced networks as shown by Eq. (4), we can get last homogeneous change framework, which decides last situation of specific finger. For instance, for pointer finger the last lattice is shown by Eq. (5).

$$A_0^{12} = T_0^1 R_x T_2^3 R_z T_5^6 R_x T_9^{10} R_x T_{11}^{12} \tag{5}$$

Some pivot grids contain straight movement also to diminish the quantity of conclusive change lattice components [15]. For assurance of finger workspace, the key is change from arrange framework 0 into the finish of the finger. Determination of conclusive change lattice is moderately basic errand and by appropriate programming should be possible.

# 3 Electrical Design

## 3.1 *Microcontroller*

We have chosen to use ATmega32U4 as our microcontroller. In addition, since Arduino is a by-and-large used progression stage, there are existing libraries open for correspondence with ROS, to be specific by methods for the ROS successive library [16].

A climb to Raspberry Pi can moreover be made if it might be fitted inside the endeavor time length. Regardless, since the Raspberry Pi just takes into consideration only one PWM GPIO, an external PWM driver should be purchased to multiplex between the servos [17]. Since ROS can legitimately run on Rasping, there will be no prerequisite for libraries to enable pass on [18].

## 3.2 *Servo Motor*

Three servos are required for the finger. One for the pointer, second one for the middle finger along with the ring finger and little finger finally the third servo for the thumb [19]. Since we found that each finger's development range can be achieved with 90-degree servos, we picked HD—3001HB servos, which considers a slowdown torque of 4.4 kg-cm at 6 V [20]. The remaining two servos are indistinguishable, and we have picked ES08MSII giving a torque of 2 kg-cm at 6 V.

## 3.3 *Power Supply*

It is noteworthy that this structure is advantageous and completely controlled by inner sources. Using a divider power supply is fine for testing and researching yet a prosthetic arm ought to be powered by a source an amputee can without a very remarkable stretch take around [21].

The muscle sensor units require close to no power; however, the servos and micro-controller require an enormous total. The muscle sensor units require two power sources to make a positive and negative voltage reference [22]. These sensors are

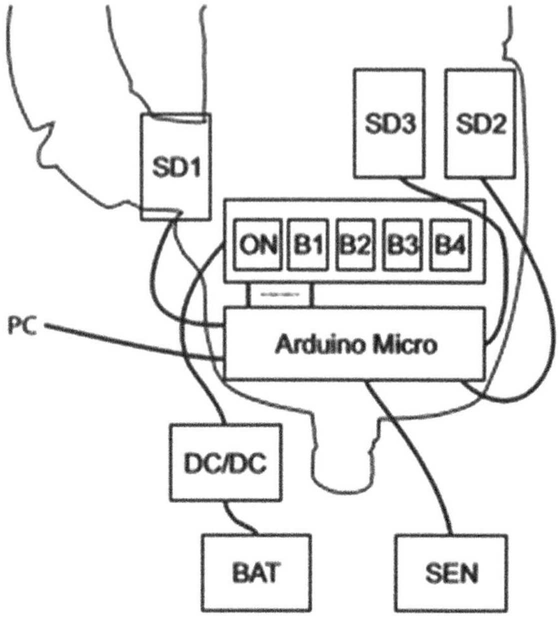


**Fig. 2** Block diagram of hand exoskeleton mechanism

unstable to enter voltage spikes and require a consistent power flexibly to create extraordinary signs.

## 3.4 *Printed Circuit Board*

Printed circuit board is the most broadly perceived name yet may in like manner be arranged "printed wiring sheets" or "printed wiring cards." As equipment created from vacuum chambers, and moves to silicon and composed circuits, on account of this advancement the size, and cost furthermore reduced [23].

The PCB has lines and pads that interface different focuses together [24]. A PCB allows signs and ability to be controlled between physical contraptions. The electrical affiliation is between the electrical fragment, and PCB is made using a patch which moreover fills in as strong mechanical cement. The block diagram of hand exoskeleton mechanism is composed of three basic elements as shown in Fig. 2.

## 3.5 *Piece*

The PCB is an elective layer of different materials that is secured using heat and a paste to shape a layer.

In a perfect world, we might want to incorporate constraint devices on each finger to give approximately input. These instruments give data to the microcontroller about how much power is being pragmatic at every fingertip [25]. This data can be utilized to control trembling engines accommodated in an adaptable band than can be worn around the upper arm. This gives approximately essential tangible input [26] to the client informing them as to whether they are getting a handle on an item and how much power they are applying. Figure 3 illustrates the CAD model of final design. The fundamental structure of the program is shown in Fig. 4.

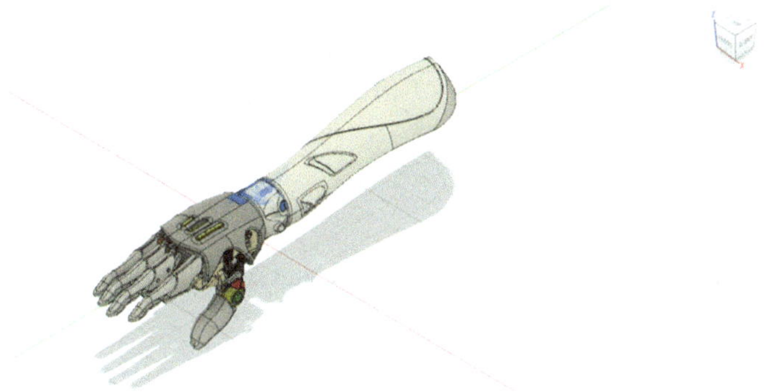

**Fig. 3** CAD model of final design (isometric view)

**Fig. 4** Fundamental structure of the program

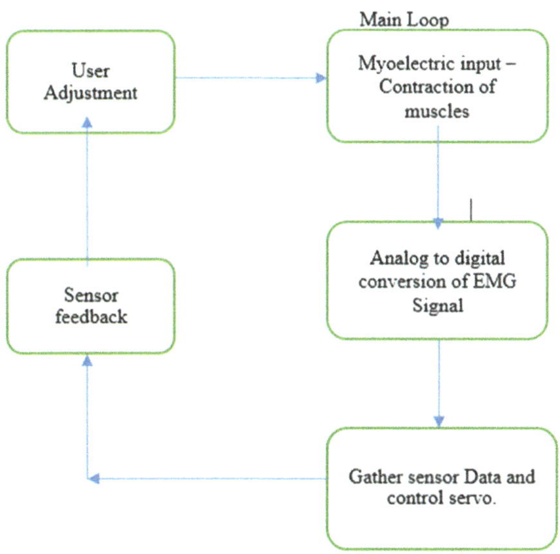

**Table 2** Overall cost of bionic arm

| S. No. | Description | Cost (₹) |
|---|---|---|
| 1 | 3D printing | 15,000 |
| 2 | Hd-3001HB servo motor X1 | 1700 |
| 3 | ES08MSII servo motor X2 | 900 |
| 4 | 7.2 V battery | 3000 |
| 5 | Arduino (for prototyping) | 2000 |
| 6 | Electronics (PCB, wire, shipping, etc.) | 9000 |
| 7 | EMG Sensor | 7000 |
| 8 | Miscellaneous (Screws, acetone, etc.) | 5000 |
| 9 | Electrode | 6950 |
| 10 | Total | 50,550 |

### *3.6 Overall Cost*

Table 2 refers the detail description of particulates and cost of the system for the development of 3D printing of prototype.

## 4 Result and Discussion

Considering the plan choices made during the fundamental structure stage, last plan was enhanced utilizing slight changes. The final prototype of electromyography-controlled bionic arm is shown in Fig. 5. An Arduino nano was picked rather than an Arduino UNO. Mowers muscle sensor was utilized. The utilization of pressure link was expelled, and an immediate connect to the fingers was built up with the servo. Explanations for these changes and the last execution steps are laid out.

The following is a picture of the printed circuit board configuration made for this prosthetic arm. A PCB schematic as shown in Fig. 6 is incredibly useful in structuring

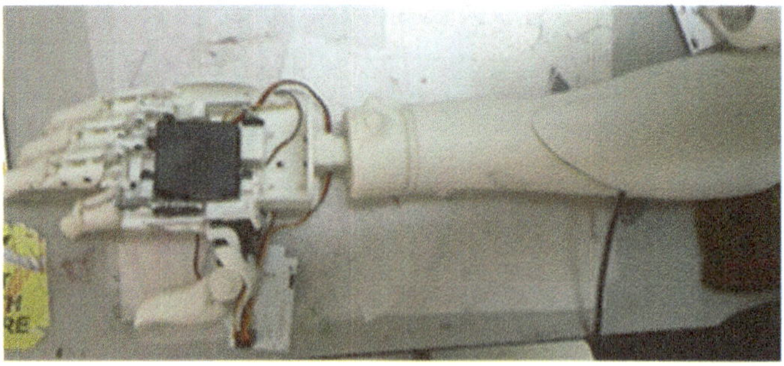

**Fig. 5** Final prototype of electromyography controlled bionic arm

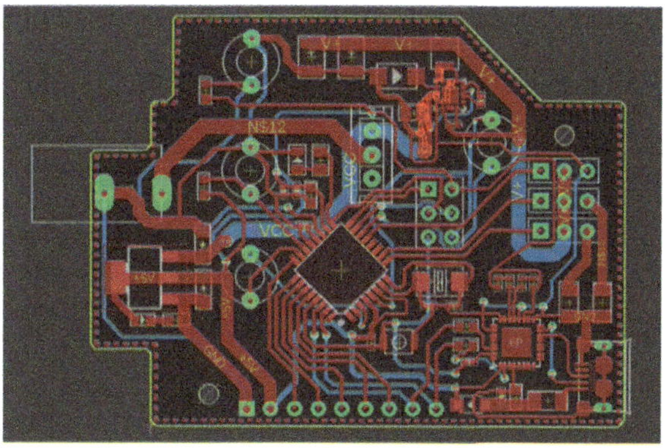

**Fig. 6** Final PCB design

model circuits on punctured EAGLE. The schematic beneath displays the servo yield pins and force supply contribution on the left followed by the voltage controllers, microcontroller, and the software development pin port for the debugger. This representation document possibly will be sent to maker to get the board manufactured as per our specifications.

## 5 Conclusions

Persuaded by the significant expense of cutting-edge humanoid automated hand, the objective of this task expected to build a reasonable 3D printed mechanical hand that can duplicate the same number of the functionalities of the propelled hand, while limiting the expense. This framework will not just permit more clients, including scientists, specialists, and amputees, who cannot bear the cost of costly propelled hand, to have the option to try and extend their use.

The framework comprises just of 3D printed parts, servos, sensors, and a smaller-scale controller. Determinations and functionalities that we wish to reproduce were characterized, and the plan that fits the rules was chosen. The small-scale controller speaks with the PC by means of sequential correspondence and facilitates control and reaction from the servos and sensors. Finally, as per the results, the prototype of the electromyography-controlled bionic arm has been developed by using the given circuit model. The developed model can be utilized as a benchmark for the further exploration of other parts in the system.

## References

1. Disselhorst-Klug C, Schmitz-Rode T, Rau G (2009) Surface electromyography and muscle force: Limits in sEMG–force relationship and new approaches for applications. Clin Biomech 24:225–235

2. Xu Z, Tian Y, Li Y (2015) sEMG pattern recognition of muscle force of upper arm for intelligent bionic limb control. J Bionic Eng 12(2):316–323
3. Clement RGE, Bugler KE, Oliver CW (2011) Bionic prosthetic hands: a review of present technology and future aspiration. Surg J Royal Colleges 9(6):336–340
4. Georgea JA, Davis TS, Brinton MR, Clark GA (2020) Intuitive neuromyoelectric control of a dexterous bionic arm using a modified Kalman filter. J Neurosci Methods 330(2):108462
5. Gundogdu K, Bayrakdar S, Yucedag I (2017) Developing and modeling of voice control system for prosthetic robot arm in medical systems. J King Saud Univ Comput Inf Sci 30(2):198–205
6. Htet HW (2016) Control system for 3d printable robotic hand. Union I Digital Works, 807 Union Street 6.
7. Pizarro YR, Schuler JM, Lippitt TC (2013) 3D printed robotic hand. Kennedy Space Center (KSC), Washington DC
8. Calì J, Calian DA, Amati C, Kleinberger R, Steed A, Kaultz J, Weyrich T (2012) 3D-printing of non-assembly, articulated models. ACM Trans Graph 31(6):130:1–130:8
9. Mazzoleni S, Sale P, Franceschini M, Bigazzi S, Carozza MC, Dario P, Posteraro F (2013) Effects of proximal and distal robot-assisted upper limb rehabilitation on chronic stroke recovery. NeuroRehabilitation 33(1):33–39
10. Gerloff C, Corewell B, Chen R, Hallett M, Cohen LG (1998) The role of the human motor cortex in the control of complex and simple finger movement sequences. Brain: J Neurol 121(9):1695–1709
11. Heo P, Gu GM, Lee S-J, Rhee K, Kim J (2012) Current hand exoskeleton technologies for rehabilitation and assistive engineering. Int J Precis Eng Manuf 13(5):807–824
12. Bos RA, Haarman CJW, Stortelder T, Nizamis K, Herder JL, Stienen AHA, Plettenburg DH (2016) A structured overview of trends and technologies used in dynamic hand orthoses. J NeuroEngineering Rehabil 13(1):1–25
13. Cesqui B, Tropea P, Micera S, Krebs HL (2013) EMG-based pattern recognition approach in post stroke robot-aided rehabilitation: a feasibility study. J Neuroengineering Rehabil 75(1):1–15
14. Song R, Tong K-Y, Hu X, Zhou W (2013) Myoelectrically controlled wrist robot for stroke rehabilitation. J NeuroEngineering Rehabil 10(1):1–8
15. Heo P, Gu GM, Lee S-J, Rhee K, Kum J (2012) Current hand exoskeleton technologies for rehabilitation and assistive engineering. Int J Precis Eng Manuf 13(5):807–824
16. Aguilar-Pereyra F, Castillo-Castaneda E (2016) Design of a reconfigurable robotic system for flexoextension fitted to hand fingers size. Appl Bionics Biomech 10(2):1–11
17. Negi S, Dhiman S, Shar RK (2014) Basics and applications of rapid prototyping medical models. Rapid Prototyping J 20(2):256–267
18. Hieu LC, Sloten JV, Hung LT, Khanh L, Soe S, Zlatov N, Phuoc LT, Trung PD (2010) Medical reverse engineering applications and methods. Proc Int Conf Innov Recent Trends Challenges Mechatron Mech Eng New High-Tech Products Dev 2(1):186–196
19. Baronio G, Harran S, Signoroni A (2016) A critical analysis of a hand orthosis reverse engineering and 3D printing process 10(2):1–8 (Hindawi Publishing Corporation)
20. Mello RGT, Oliveira LF, Nadal J (2007) Digital Butterworth filter for subtracting noise from low magnitude surface electromyogram. Comput Methods Programs Biomed 87(1):28–35
21. Yeow C-H, Baisch AT, Talbot SG, Walsh CJ (2014) Cable-driven finger exercise device with extension return springs for recreating standard therapy exercises. J Med Dev 8(2):014502
22. De Luca CJ, Gilmore LD, Kuznetsov M, Roy SH (2010) Filtering the surface EMG signal: movement artifact and baseline noise contamination. J Biomech 43(2):1573–1579
23. Phinyomark A, Phukpattaranont P, Limsakul C (2012) Feature reduction and selection for EMG signal classification. Expert Syst Appl 39(2):7420–7431
24. Oskoei MA, Hu H (2008) Support vector machine-based classification scheme for myoelectric control applied to upper limb. IEEE Trans Biomed Eng 55(8):1956–1965

25. Abdallah IB, Bouteraa Y, Rekik C, (2017) Design and development of 3D printed myoelectric robotic exoskeleton for hand rehabilitation. Int J Smart Sensing Intell Syst 10(2)
26. Canizares A, Pazos J, Benitez D (2017) On the use of 3D printing technology towards the development of a low-cost robotic prosthetic arm. IEEE Int Autumn Meeting Power Electron Comput 15(2):1–6

# Multi-response Optimization of Machining Characteristics Using MQL Through GRA and TOPSIS Approach

**A. Venkata Vishnu, S. Sudhakar Babu, and P. Jamaleswara Kumar**

**Abstract** In machining cutting, fluids play a vital role in terms of tool life, cutting temperature, surface finish, etc., due to its functions like continuous lubrication, cooling and chips flushing. Nowadays, advanced cutting fluids like nanofluids are effectively used in machining due to its heat transfer medium at the cutting zone, and on the other hand, vegetable oil-based cutting fluids which are eco-friendly are being widely used as coolants in different machining operations. However, these conventional cutting fluids are difficult to process and are very costly. In order to eliminate disaffects associated with the cutting fluids, it is suggested to shift towards minimum quantity lubrication/near dry machining. In the present work, the performance characteristics of vegetable oil and nanofluids in machining of EN353 alloy steel under MQL conditions are compared. Using Taguchi grey relational analysis, an optimum solution is obtained for the output responses, i.e. cutting temperature, surface roughness (Ra and Rz) and material removal rate. It was concluded that, vegetable oil showed better performance compared to nanofluid ($Al_2O_3$).

**Keywords** Minimum quantity of lubrication (MQL) · Nanofluids · Vegetable oil-based fluids · Cutting temperature · Surface roughness · Material removal rate · etc.

## 1 Introduction

According to Bruni et al. [1], machining plays an important role in converting raw material to a desired shape by metal removal in the form of chips; lot of heat is generated near tool and workpiece interface due to the development of friction between them, where cutting fluids are employed to overcome this effect. Cutting fluids play an important role by cooling tool and work piece interface and flushing the chips away from the cutting zone. Khan et al. [2] and Debnath et al. [3] have reviewed the machining performance of vegetable-based cutting fluids compared to mineral-based

A. Venkata Vishnu (✉) · S. S. Babu · P. J. Kumar
Department of Mechanical Engineering, Koneru Lakshmaiah Education Foundation, Guntur, Andhra Pradesh, India



P. Srinivasa Pai and V. Krishnaraj (eds.), *Sustainable Machining Strategies for Better Performance*, Lecture Notes in Mechanical Engineering,
https://doi.org/10.1007/978-981-16-2278-6_3

cutting fluids which significantly reduced the ecological problems. Vasu et al. [4], Pham et al. [5] and Gupta et al. [6] evaluated the performance characteristics like tool life, surface finish and cutting temperatures using nanoparticle-enhanced cutting fluids in machining where considerable improvement has observed.

According to Khan et al. [2], Sharma et al. [7], cutting fluids account around 16 to 20% of the total manufacturing costs. Among various techniques available on the reduction of coolant flow in machining, Amini et al. [8], Ramana et al. [9], Vishnu [10–12] suggest near dry machining (NDM)/minimum quantity lubrication (MQL) as a viable alternate, which is a sustainable manufacturing technique minimizes the use of coolant flow by mixing coolant with air. The lubricant is mixed with air, and a mist is formed and sprayed near the cutting zone.

Ramana et al. [9] implemented the Taguchi methodology successfully though selection of orthogonal array depending on total number of factors and its levels.

As per the recent advancements, machining using nanofluids and vegetable oil-based lubricants is gaining more advantages compared to other conventional fluids. In the present work, a comparison is made between lubricants of vegetable oil and nanofluids in machining under MQL condition. The work material selected for machining is EN 353 alloy steel with carbon content of 0.17% for its applications in manufacturing of gears, shafts, pinions, camshafts and gudgeon pins, etc. [13]. The performance characteristics of surface roughness (Ra and Rz), cutting temperature and material removal rate are studied by considering input parameters at three different levels. Taguchi grey relational analysis and TOPSIS method are used to study the performance behaviour of input parameters with respect to output parameters.

## 2 Experimentation

In this paper, the experiments are performed considering standard $L_9$ ($3^4$) orthogonal array with the input parameters; i.e. speed, feed, depth of cut and type of tool at three different levels were given in Tables 1 and 2.

The experimentation is carried out at SR INDUSTRIES, Hyderabad, and the CNC machine used is LOKESH TL20 Max CNC MACHINE. Nine specimens for 9 experiments of EN 353 alloy steel with 35 mm Dia and 150 mm long are considered

**Table 1** Control factors with levels

| Factors/levels | Speed (rpm) (A) | Feed (mm/rev) (B) | Depth of cut (mm) (C) | Type of tool (D) |
|---|---|---|---|---|
| 1 | 700 | 0.2 | 0.5 | T-I |
| 2 | 1100 | 0.5 | 1.5 | T-II |
| 3 | 1500 | 0.8 | 2.5 | T-III |

**Table 2** Standard $L_9$ $(3^4)$ orthogonal array with input parameters

| Experiment No. | A | B | C | D |
|---|---|---|---|---|
| 1 | 700 | 0.2 | 0.5 | T-I |
| 2 | 700 | 0.5 | 1.5 | T-II |
| 3 | 700 | 0.8 | 2.5 | T-III |
| 4 | 1100 | 0.2 | 1.5 | T-III |
| 5 | 1100 | 0.5 | 2.5 | T-I |
| 6 | 1100 | 0.8 | 0.5 | T-II |
| 7 | 1500 | 0.2 | 2.5 | T-II |
| 8 | 1500 | 0.5 | 0.5 | T-III |
| 9 | 1500 | 0.8 | 1.5 | T-I |

for machining. The cutting tools used for machining are TNMG carbide tool of uncoated (T-I), CVD (T-II) and PVD (T-III) inserts (Fig. 1).

The coolants used are vegetable oil-based fluid which is prepared by mixing 40% of coconut oil + 40% of oleic acid + 20% of triethanolamine and stirred properly, and on the other end, nanofluids were prepared by mixing and stirring $Al_2O_3$ nano-sized particles with base fluid water and stabilizers. A total of nine experiments were performed as per the orthogonal array using vegetable oil and nanofluid separately

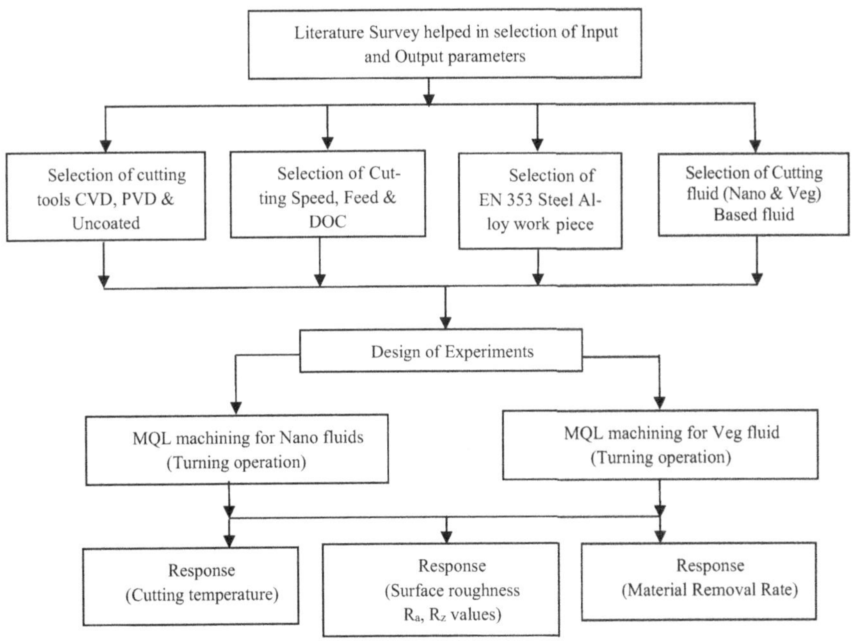


**Fig. 1** Schematic layout of experimentation

and the output parameters—cutting temperature, surface roughness (Ra and Rz), and material removal rate were measured. Surface roughness is measured using a portable surface roughness tester MITUTOYO, and the results of corresponding Ra and Rz values are given in Table 3. While machining, cutting temperature is measured using portable temperature measuring device. The material removal rate is also calculated by using Eq. (1), and the weights (before and after machining) are measured using standard weighing machine.

$$\mathrm{MRR} = (1000 \times W)/(\rho \times t)\ \mathrm{mm}^3/\mathrm{min} \quad (1)$$

$W = \mathrm{Wi} - \mathrm{Wf}$
Wi= Initial weight—Before machining weight in gms
Wf= Final weight—After machining weight in gms
$\rho$ = Density of EN 353 steel alloy (8.08 g/cm$^3$)
$t$ = Machining time in min.

The MQL setup as shown in Fig. 2 is developed with the help of a paint zoom consisting of an air compressor, where the air pressure is maintained at a pressure

**Table 3** Surface roughness, cutting temperature and material removal rate values

| Experiment No. | Vegetable oil | | | | Nanofluid | | | |
|---|---|---|---|---|---|---|---|---|
| | Cutting temperature | Ra | Rz | MRR | Cutting temperature | Ra | Rz | MRR |
| 1 | 27.05 | 3.25 | 17.275 | 518.4 | 28.15 | 3.465 | 18.545 | 354.2 |
| 2 | 28.215 | 3.57 | 18.07 | 5934.05 | 37.6 | 3.69 | 19.11 | 4748.12 |
| 3 | 31.68 | 3.77 | 18.505 | 11,729.525 | 32.1 | 3.765 | 19.5 | 10,775.7 |
| 4 | 28.45 | 3.39 | 17.7 | 4695.36 | 30.15 | 3.395 | 18.34 | 3961.05 |
| 5 | 32.35 | 3.505 | 18.885 | 7947.37 | 30.9 | 3.61 | 19.07 | 7957.25 |
| 6 | 31.1 | 3.485 | 17.935 | 1060.84 | 30.1 | 3.51 | 18.88 | 628.45 |
| 7 | 32.4 | 3.64 | 18.215 | 6609.7 | 31.4 | 3.67 | 19.295 | 6618.545 |
| 8 | 29.365 | 3.565 | 18.01 | 551.55 | 29.2 | 3.57 | 18.95 | 649.66 |
| 9 | 33.05 | 3.455 | 17.115 | 9046.95 | 39.15 | 3.39 | 18.615 | 7036.1 |

**Fig. 2** MQL setup

**Fig. 3** MQL machining

of 1 bar; on the other end a coolant inlet maintaining a flow rate of 150 ml/h mixed together and sprayed near the cutting zone with the help of a nozzle as shown in Fig. 3, the results obtained are given in Table 3 and are compared among vegetable oil and nanofluid coolant conditions.

# 3 Results and Discussions

## 3.1 *Optimization of Input Parameters*

To investigate the statistical significance of the input parameters that affect the machinability of alloy steel, ANOVA was performed. Tables 4, 5, 6 and 7 show the results of ANOVA with the cutting temperature, surface roughness (Ra and Rz) and MRR for vegetable oil and nanofluids.

The analysis was carried out at a level of significance of 0.05% and for the level of confidence of 99.95%. From the tables, it was observed that all the input parameters are significant on comparing the *F*-values with the tabulated ones, where all the *F*

**Table 4** ANOVA results for cutting temperature

| Vegetable oil | | | | | | Nanofluid | | | | |
|---|---|---|---|---|---|---|---|---|---|---|
| Source | S.S | DOF | M.S.S | F ratio | $\rho\%$ | S.S | DOF | M.S.S | F ratio | $\rho\%$ |
| A | 21.10 | 2 | 10.55 | 116.73 | 27.66 | 27.21 | 2 | 13.61 | 66.02 | 11.79 |
| B | 22.62 | 2 | 11.31 | 125.13 | 29.67 | 47.34 | 2 | 23.67 | 114.85 | 20.64 |
| C | 28.75 | 2 | 14.38 | 159.04 | 37.78 | 129.52 | 2 | 64.76 | 314.21 | 56.78 |
| D | 3.15 | 2 | 1.58 | 17.45 | 3.93 | 23.31 | 2 | 11.65 | 56.55 | 10.07 |
| Error | 0.81 | 9 | 0.09 | | 0.96 | 1.855 | 9 | 0.21 | | 0.73 |
| Mean | 16,642.18 | 1 | | | | 18,528.12 | 1 | | | |
| Total | 16,718.64 | 18 | | | | 18,757.37 | 18 | | | |

**Table 5** ANOVA results for surface roughness (Ra)

| Vegetable oil | | | | | | Nanofluid | | | | |
|---|---|---|---|---|---|---|---|---|---|---|
| Source | S.S | DOF | M.S.S | F ratio | $\rho$% | S.S | DOF | M.S.S | F ratio | $\rho$% |
| A | 0.0283 | 2 | 0.014 | 5.221 | 6.48 | 0.058 | 2 | 0.029 | 14.724 | 19.16 |
| B | 0.070 | 2 | 0.035 | 13.090 | 18.55 | 0.039 | 2 | 0.020 | 9.907 | 12.44 |
| C | 0.142 | 2 | 0.071 | 26.289 | 38.81 | 0.128 | 2 | 0.064 | 32.665 | 44.22 |
| D | 0.11 | 2 | 0.056 | 20.547 | 30.00 | 0.056 | 2 | 0.028 | 14.301 | 18.57 |
| Error | 0.024 | 9 | 0.003 | | 6.14 | 0.017 | 9 | 0.002 | | 5.58 |
| Mean | 222.32 | 1 | | | | 228.48 | 1 | | | |
| Total | 222.70 | 18 | | | | 228.78 | 18 | | | |

calculated values are greater than the tabulated value which was 4.26. From Table 8, it was also observed that depth of cut has more per cent of contribution compared to all other input parameters under vegetable and nanofluid conditions for cutting temperature, surface roughness (Ra and Rz) and MRR.

## 3.2 *Multi-response Optimization Through GRG*

The objective of the present paper is to identify the performance characteristics between vegetable oil and nanofluids coolants using MQL. As the output, parameter is not a single response. Multiple responses of optimization of surface roughness (Ra and Rz), cutting temperature and material removal rate are obtained using Taguchi grey relational analysis [14–18] for both vegetable oil and nanofluid results individually. The grey relational analysis was employed to determine the optimum combination of turning parameters, i.e. speed, feed, depth of cut and type of tool that minimizes the responses cutting temperature and surface roughness Ra and Rz (smaller the better), maximizes material removal rate (larger the better). The normalized and grey relational coefficient of cutting temperature, surface roughness (Ra and Rz values) and the material removal rate values were calculated accordingly and given in Tables 9 and 10 under MQL condition for vegetable oil and nanofluid. The highest value of GRG obtained through grey relational coefficient is considered as the stronger relational degree. From Table 9, the highest GRG for vegetable oil is 0.787 which is experiment no 1; similarly from Table 10 for nanofluid, 0.768 is the highest.

The grey relational grade (GRG) response for vegetable oil and nanofluid is given in Table 11. The highest average value of GRG represents best result. The optimal level of each factor is determined using the results of GRG. Table 11 shows the optimum levels for machining with vegetable oil, i.e. A1, B1, C2 and D1, with type of tool as the most significant factor. Similarly, for nanofluids A2, B1, C1 and D1 with type of tool as the most significant factor, the predicted responses for cutting

**Table 6** ANOVA results for surface roughness (Rz)

| Vegetable oil | | | | | | Nanofluid | | | | |
|---|---|---|---|---|---|---|---|---|---|---|
| Source | S.S | DOF | M.S.S | F ratio | $\rho\%$ | S.S | DOF | M.S.S | F ratio | $\rho\%$ |
| A | 0.466978 | 2 | 0.23 | 41.45 | 9.18 | 0.257811111 | 2 | 0.13 | 4.89 | 9.25 |
| B | 1.171544 | 2 | 0.59 | 103.98 | 23.36 | 0.352211111 | 2 | 0.18 | 6.68 | 13.51 |
| C | 2.93378 | 2 | 1.47 | 260.33 | 58.83 | 1.234711111 | 2 | 0.62 | 23.41 | 53.33 |
| D | 0.394811 | 2 | 0.20 | 35.04 | 7.72 | 0.371477778 | 2 | 0.19 | 7.04 | 14.38 |
| Error | 0.0507 | 9 | 0.01 | | 0.91 | 0.23735 | 9 | 0.03 | | 9.52 |
| Mean | 5811.139 | 1 | | | | 6445.287339 | 1 | | | |
| Total | 5816.156 | 18 | | | | 6447.7409 | 18 | | | |

**Table 7** ANOVA results for material removal rate

| Vegetable oil | | | | | | Nanofluid | | | | |
|---|---|---|---|---|---|---|---|---|---|---|
| Source | S.S | DOF | M.S.S | F ratio | $\rho$% | S.S | DOF | M.S.S | F ratio | $\rho$% |
| A | 6,716,682 | 2 | 3,358,341 | 6062 | 2.62 | 3,702,875 | 2 | 1,851,437 | 93,314.53 | 1.70 |
| B | 3,598,025 | 2 | 17,990,128 | 3247 | 14.03 | 19,570,824 | 2 | 9,785,412 | 493,195 | 8.97 |
| C | 20,778,781 | 2 | 103,893,905 | 187,556 | 81.02 | 18,978,998 | 2 | 94,894,992.86 | 4,782,814 | 87.01 |
| D | 5,984,537 | 2 | 2,992,268 | 5401 | 2.33 | 5,053,602 | 2 | 2,526,801 | 12,735 | 2.32 |
| Error | 4985.3 | 9 | 553.93 | | 0.00 | 178.56 | 9 | 19.84 | | 0.00 |
| Mean | 5,140,018 | 1 | | | | 4,057,275 | 1 | | | |
| Total | 7,704,761 | 18 | | | | 6,238,449 | 18 | | | |

**Table 8** Percent contribution factor for vegetable oil and nanofluid parameters

| Vegetable oil | | | | | Nanofluid | | | |
|---|---|---|---|---|---|---|---|---|
| Source | Cutting temperature | Ra | Rz | MRR | Cutting temperature | Ra | Rz | MRR |
| A | 27.66 | 6.48 | 9.18 | 2.62 | 11.79 | 19.16 | 9.25 | 1.70 |
| B | 29.67 | 18.55 | 23.36 | 14.03 | 20.64 | 12.44 | 13.51 | 8.97 |
| C | 37.78 | 38.81 | 58.83 | 81.02 | 56.78 | 44.22 | 53.33 | 87.01 |
| D | 3.93 | 30.00 | 7.72 | 2.33 | 10.07 | 18.57 | 14.38 | 2.32 |
| Error | 0.96 | 6.14 | 0.91 | 0.00 | 0.73 | 5.58 | 9.52 | 0.00 |

temperature, surface roughness (Ra and Rz) and MRR were estimated by Eq. (2) for vegetable oil and Eq. (3) for nanofluid, where *A*, *B*, *C* and *D* are the corresponding input parameters, $Y_{ij}$ is the average response value for all the nine experiments.

$$(\text{Predicted Response})_{\text{veg oil}} = \text{A1} + \text{B1} + \text{C2} + \text{D1} - 3 * \left(Y_{ij}\right) \tag{2}$$

$$(\text{Predicted Response})_{\text{nano fluid}} = \text{A2} + \text{B1} + \text{C1} + \text{D1} - 3 * \left(Y_{ij}\right) \tag{3}$$

### *3.3 Multi-response Optimization Through TOPSIS*

The TOPSIS-based optimization was performed, and the detailed steps of TOPSIS are referred [19, 20]. The closeness coefficients (CC) are measured considering the performance criterion, ideal best distance and ideal worst distance which is given in Tables 12 and 13 for vegetable oil and nanofluid.

## 4 Conclusions

In the present work, a comparison is made between lubricants of vegetable oil and nanofluids in CNC machining of alloy steel using MQL technique, by considering the input parameters. Taguchi grey relational analysis and TOPSIS method are used to analyse and validate the data for optimum solution. From Fig. 4, the larger the better S/N ratios is plotted for GRG values of both vegetable oil and nanofluid of which vegetable oil results are maximum and showed favourable results compared to nanofluids.

**Table 9** GRG of vegetable oil

| Exp. No. | Normalized values | | | | Grey relational coefficient | | | | GRG | S/N rations | Rank |
|---|---|---|---|---|---|---|---|---|---|---|---|
| | Cut. Temp. | $R_A$ | $R_Z$ | MRR | Cut. Temp. | $R_A$ | $R_Z$ | MRR | | | |
| 1 | 1.000 | 1.000 | 0.885 | 0.000 | 1.000 | 1.000 | 0.813 | 0.333 | 0.787 | 2.081 | 1 |
| 2 | 0.778 | 0.385 | 0.313 | 0.483 | 0.693 | 0.448 | 0.421 | 0.492 | 0.513 | 5.798 | 5 |
| 3 | 0.223 | 0.000 | 0.000 | 1.000 | 0.392 | 0.333 | 0.333 | 1.000 | 0.515 | 5.764 | 4 |
| 4 | 0.700 | 0.731 | 0.579 | 0.373 | 0.625 | 0.650 | 0.543 | 0.443 | 0.565 | 4.959 | 3 |
| 5 | 0.067 | 0.510 | 0.446 | 0.663 | 0.349 | 0.505 | 0.474 | 0.597 | 0.481 | 6.357 | 6 |
| 6 | 0.317 | 0.548 | 0.410 | 0.048 | 0.423 | 0.525 | 0.459 | 0.344 | 0.438 | 7.171 | 8 |
| 7 | 0.100 | 0.250 | 0.209 | 0.543 | 0.357 | 0.400 | 0.387 | 0.523 | 0.417 | 7.597 | 9 |
| 8 | 0.612 | 0.394 | 0.356 | 0.003 | 0.563 | 0.452 | 0.437 | 0.334 | 0.447 | 6.994 | 7 |
| 9 | 0.000 | 0.606 | 1.000 | 0.761 | 0.333 | 0.559 | 1.000 | 0.676 | 0.642 | 3.849 | 2 |

**Table 10** GRG of nanofluid

| Exp. No. | Normalized values | | | | Grey relational coefficient | | | | GRG | S/N rations | Rank |
|---|---|---|---|---|---|---|---|---|---|---|---|
| | Cut. Temp. | $R_A$ | $R_Z$ | MRR | Cut. Temp. | $R_A$ | $R_Z$ | MRR | | | |
| 1 | 1.000 | 1.000 | 0.823 | 0.000 | 1.000 | 1.000 | 0.739 | 0.333 | 0.768 | 2.293 | 1 |
| 2 | 0.114 | 0.161 | 0.336 | 0.414 | 0.361 | 0.373 | 0.430 | 0.461 | 0.406 | 7.829 | 9 |
| 3 | 0.667 | 0.000 | 0.000 | 1.000 | 0.600 | 0.333 | 0.333 | 1.000 | 0.567 | 4.928 | 4 |
| 4 | 0.810 | 0.806 | 1.000 | 0.338 | 0.724 | 0.721 | 1.000 | 0.430 | 0.719 | 2.865 | 2 |
| 5 | 0.800 | 0.333 | 0.371 | 0.726 | 0.714 | 0.429 | 0.443 | 0.646 | 0.558 | 5.067 | 5 |
| 6 | 0.838 | 0.548 | 0.534 | 0.014 | 0.755 | 0.525 | 0.518 | 0.336 | 0.534 | 5.449 | 7 |
| 7 | 0.733 | 0.204 | 0.177 | 0.596 | 0.652 | 0.386 | 0.378 | 0.553 | 0.492 | 6.161 | 8 |
| 8 | 0.933 | 0.419 | 0.474 | 0.016 | 0.882 | 0.463 | 0.487 | 0.337 | 0.542 | 5.320 | 6 |
| 9 | 0.000 | 0.806 | 0.763 | 0.637 | 0.333 | 0.721 | 0.678 | 0.579 | 0.578 | 4.761 | 3 |

**Table 11** GRG response table

| Veg oil | | | | | | Nanofluid | | | | |
|---|---|---|---|---|---|---|---|---|---|---|
| Factors/levels | 1 | 2 | 3 | Max-Min | Rank | 1 | 2 | 3 | Max-Min | Rank |
| Speed | **0.605** | 0.494 | 0.502 | 0.111 | 2 | 0.580 | **0.604** | 0.537 | 0.067 | 4 |
| feed | **0.589** | 0.480 | 0.531 | 0.109 | 3 | **0.660** | 0.502 | 0.560 | 0.158 | 2 |
| Depth of cut | 0.557 | **0.573** | 0.471 | 0.102 | 4 | **0.615** | 0.568 | 0.539 | 0.076 | 3 |
| Type of tool | **0.636** | 0.456 | 0.509 | 0.18 | 1 | **0.635** | 0.477 | 0.609 | 0.158 | 1 |

The bolded values are optimum values for the experiment

**Table 12** CC of vegetable oil

| Si+ | Si− | CC | Rank |
|---|---|---|---|
| 0.0007 | 0.0212 | 0.9660 | 1 |
| 0.0095 | 0.0137 | 0.5904 | 3 |
| 0.0188 | 0.0036 | 0.1622 | 9 |
| 0.0065 | 0.0150 | 0.6976 | 2 |
| 0.0168 | 0.0070 | 0.2944 | 7 |
| 0.0130 | 0.0089 | 0.4059 | 5 |
| 0.0181 | 0.0037 | 0.1712 | 8 |
| 0.0106 | 0.0113 | 0.5158 | 4 |
| 0.0170 | 0.0099 | 0.3673 | 6 |

**Table 13** CC of nanofluid

| Si+ | Si− | CC | Rank |
|---|---|---|---|
| 0.001 | 0.029 | 0.970 | 1 |
| 0.026 | 0.004 | 0.133 | 9 |
| 0.015 | 0.018 | 0.544 | 7 |
| 0.006 | 0.024 | 0.813 | 2 |
| 0.010 | 0.022 | 0.695 | 5 |
| 0.007 | 0.023 | 0.771 | 4 |
| 0.012 | 0.020 | 0.623 | 6 |
| 0.007 | 0.026 | 0.783 | 3 |
| 0.027 | 0.010 | 0.263 | 8 |

Grey relational analysis is employed successfully where it was observed that there was an improvement in the confirmatory results and GRG compared with the initial and predicted values for both vegetable and nanofluids as shown in Table 14. The optimization for multi-responses using grey relational analysis is compared with TOPSIS method where the generated results are shown in Tables 9 and 12 for

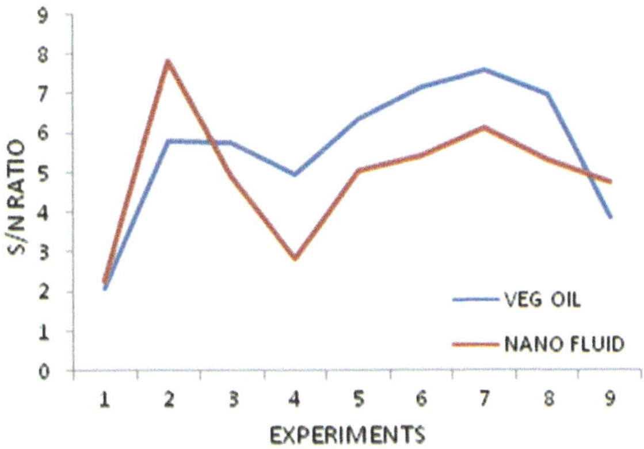


**Fig. 4** S/N ratio plot for vegetable oil and nanofluid

vegetable oil and Tables 10 and 13 for nanofluid. It is observed that machining with vegetable oil, the highest GRG value obtained through grey relational analysis was 0.787, the ranking results of GRA is compared with ranking results of TOPSIS for which the highest closeness coefficient value is 0.9660 for experiment no. 1. For nanofluid under the highest GRG value obtained through grey relational analysis was 0.768, the ranking results of GRA are compared with ranking results of TOPSIS for which the highest closeness coefficient value is 0.97016 for experiment no. 1. As it was observed that the first experiment sequence is found to be the closest optimum solution when compared with both the optimization techniques.

Table 14 Confirmatory results

| Veg oil | | | | Nanofluid | | |
|---|---|---|---|---|---|---|
| | Initial cutting parameters | Optimum parameters | | Initial cutting parameters | Optimum parameters | |
| | | Predicted values | Experiment values | | Predicted values | Experiment values |
| Level | A1,B1,C1,D1 | A1,B1,C2,D1 | A1,B,C2,D1 | A1,B1,C1,D1 | A2,B1,C1,D1 | A2,B1,C1,D1 |
| Cutting temperature | 27.05 | | 25.6 | 28.15 | | 26.8 |
| Ra | 3.25 | | 3.05 | 3.465 | | 3.1 |
| Rz | 17.275 | | 17.01 | 18.545 | | 17.69 |
| MRR | 518.4 | | 6869.2 | 354.2 | | 5471.3 |
| GRG | 0.787 | 0.804 | 0.884 | 0.768 | 0.795 | 0.872 |

## References


1. Bruni C, Gabrielli FF, Simoncini M (2006) Effect of the lubrication-cooling technique, insert technology and machine bed material on the work part surface finish and tool wear in finish turning of AISI 420B. Int J Mach Tools Manuf 46:1547–1554
2. Khan MMA, Mithu MAH, Dhar NR (2009) Effects of minimum quantity lubrication on turning AISI 9310 alloy steel using vegetable oil based cutting fluid. J Mater Process Technol 209:5573–5583
3. Debnath S, Reddy MM, Yi QS (2014) Environmental friendly cutting fluids and cooling techniques in machining: a review. J Cleaner Prod 83:33–47
4. Vasu V, Reddy PKG (2011) Effect of minimum quantity lubrication with $Al_2O_3$ nanoparticles on surface roughness, tool wear and temperature dissipation in machining Inconel 600 alloy. Proc Inst Mech Eng Part N J Nanoeng Nanosyst 225:3–16
5. Pham MQ, Yoon HS, Khare V, Ahn SH (2014) Evaluation of ionic liquids as lubricants in micro milling-process capability and sustainability. J Cleaner Prod 76:167–173
6. Gupta MK, Sood PK, Sharma VS (2016) Optimization of machining parameters and cutting fluids during nano-fluid based minimum quantity lubrication turning of titanium alloy by using evolutionary techniques. J Cleaner Prod 135
7. Sharma AK, Tiwari AK, Dixit AR (2016) Effects of minimum quantity lubrication (MQL) in machining processes using conventional and nanofluid based cutting fluids: a comprehensive review. J Cleaner Prod 127:1–18
8. Amini S, Khakbaz H, Barani A (2015) Improvement of near-dry machining and its effect on tool wear in turning of AISI 4142. Mater Manuf Process 30:241–247. ISSN: 1042-6914 print: 1532-2475
9. Venkata Ramana M, Krishna Mohan Rao G, Hanumantha Rao D Experimental investigations on machining of titanium alloy under different machining environments. In: Recent Advances in Mechanical Engineering Applications, pp 137–145. ISBN: 978-960-474-345-2
10. Venkata Vishnu A, Jamaleswara Kumar P, Venkata Ramana M (2018) Comparison among dry, flooded and MQL conditions in machining of EN 353 steel alloys-an experimental investigation. Mater Today: Proc
11. Vishnu AV, Kumar PJ (2019) Experimental investigations in turning of steel alloy under different lubrication conditions. In: Recent advances in material sciences. Springer
12. Vishnu AV, Kumar PJ (2018) Investigations on high speed machining of EN-353 steel alloy under different machining environments. IOP Conf Ser: Mater Sci Eng 330:012085
13. Emmanuel J, Davis R (2013) An experimental study and analysis of surface roughness in wet turning operation of EN 353 steel. Int J Current Eng Technol 3(3). ISSN 2277-4106
14. Ross PJ (2005) Taguchi techniques for quality engineering, 2nd edn. Tata McGraw Hill
15. Sivaiah P, Chakradhar D (2018) Performance improvement of cryogenic turning process during machining of 17–4 PH stainless steel using multi objective optimization techniques. Measurement. https://doi.org/10.1016/j.measurement.2018.12.094
16. Kanchana J, Prasath V, Geetha Priyadarshini B (2019) Multi response optimization of process parameters using grey relational analysis for milling of hardened custom 465 steel. Procedia Manuf 30:451–458
17. Aravind S, Shunmugesh K, Biju J, Vijayan JK (2017) Optimization of micro-drilling parameters by Taguchi grey relational analysis. Mater Today: Proc 4:4188–4195
18. Sarikaya M, Gullu A (2014) Multi-response optimization of MQL parameters using Taguchi-based GRA in turning of difficult to cut alloy Haynes 25. J Cleaner Prod 1–11
19. Parthiban K, Duraiselvam M, Manivannan R (2018) TOPSIS based parametric optimization of laser micro-drilling of TBC coated nickel based superalloy. Opt Laser Technol 102:32–39
20. Naik SS, Rana J, Nanda P (2018) Using TOPSIS method to optimize the process parameters of D2 steel on electro-discharge machining. Int J Mech Eng Technol 9(13):1083–1090

# An Experimental Investigation of Laser-Assisted Machining of EN24 Steel

**Ajit M. Hebbale, S. Rajesh K. Reddy, Mirza Abdul Hadi Baig, Manish Tak, and Ravi N. Bathe**

**Abstract** The demand for higher strength and heat-resistant materials is increasing day by day. These materials are considered as difficult to machine due to excessive tool wear and poor surface finish. EN24 is commonly used steel in the aerospace industry where machinability for this material is limited due to their higher strength and hardness. In the present study, the laser-assisted machining [LAM] process of EN24 steel was studied using a traditional lathe system integrated with a 6 kW fiber delivered diode laser. A universal 5-axes fixture was developed to mount the optical unit on the lathe machine to facilitate manual alignment and synchronized movement of laser spot with the tool. A systematic laser-assisted machining study was carried out, and surface roughness and tool wear were examined by varying parameters of laser power, spindle speed, and depth of cut. A three-dimensional transient, heat transfer model of the process was developed through the COMSOL Multiphysics software tool to compute temperature profile in the workpiece during LAM. The experimental results showed significant improvement in machinability and reduction in tool wear in the LAM process.

**Keywords** LAM · Surface roughness · Tool life · COMSOL

## 1 Introduction

The industry needs components that have a high load-carrying capacity, high stiffness, durability, good tempering properties, resistance to corrosion, fatigue, wear, and can work under extreme loading conditions, which includes aerospace and automotive, electricity, tool and die, metal machinery, etc. These applications need the use of greater material strengths like tool steel, stainless steel, and hardened steel,

A. M. Hebbale (✉) · S. R. K. Reddy · M. A. H. Baig · M. Tak · R. N. Bathe
International Advanced Research Centre for Powder Metallurgy and New Materials (ARCI), Hyderabad 500005, India

A. M. Hebbale
Department of Mechanical Engineering, NMAM Institute of Technology, Nitte, Karkala, Karnataka 574110, India



P. Srinivasa Pai and V. Krishnaraj (eds.), *Sustainable Machining Strategies for Better Performance*, Lecture Notes in Mechanical Engineering,
https://doi.org/10.1007/978-981-16-2278-6_4

and other super alloys with specific metallurgical properties to meet the demands of extreme applications. On the other side, materials that are tougher and harder are very difficult to machine. Manufacturing industries are aggressively seeking lower-cost alternatives, including innovative methods for processing these high-strength materials while meeting efficiency, safety, and a decrease in energy waste requirements [1].

Conventional machining techniques such as hard turning have gained wide industrial acceptance due to the greater material removal rates that are possible with the use of wear-resistant tools, hard and the better quality of machined surfaces [1, 2]. The hard turning technology developments can be broadly categorized as (a) improving the quality of cutting tool materials and (b) hot machining. In the early classified stages, the beginning preparation of special cutting edge tool and tough ceramic tools has allowed longer life during the machining of difficult to machine materials. On the other hand, hot machining has become critical for machining materials that are difficult to machine because they minimize the stiffness of the region of material to be extracted without affecting the bulk material. It is been well-known that there is a decrease in the flow stress and strain-hardening rate of the material with an increase in temperature. Since the 1960s, machining with the aid of a laser beam has been studied as one among the many techniques for imparting heat to the working material [2]. In the last two decades, $CO_2$ and Nd: YAG high-power lasers have become well-accepted as machining instruments in industrial production. A comparatively higher operating costs and complexity of $CO_2$ and Nd: YAG laser systems were resolved by the introduction of high-power diode lasers that combine high efficiency with a compact design that can be used in multiple applications because of their power, spectral and beam quality, wavelength. In addition, these diode lasers can be controlled in terms of positioning and operation time; therefore, diode lasers have become a flexible tool for several manufacturing processes [3]. In the recent past, a new technique LAM process is introduced where the material is precisely heated just before the removal of the material by a tool. The laser is focused on the workpiece as a small beam where the material is heated at a controlled temperature just above the tool and the temperature is controlled in such a way that at that point elevated temperature the materials ultimate strength and yield strength decreases or in other words it becomes softer during machining [4]. Thus, laser-assisted turning comes into the picture where machining is done on the cylindrical workpiece. LAM can achieve lower cutting forces, slower tool wear progression, higher material removal rate, and better surface quality. Precise control of temperature is essential to implement the LAM process. LAM studies have shown that operation parameters such as laser power, laser lead distance to the tool, cutting speed, and feed can greatly affect the material temperature in the cutting zone and the resulting cutting forces. Chip morphology is an important attribute in the cutting process [5, 6]. Gratias et al. [7] reported about the LAM process and observed that the laser power and cutting speed relations of machining of hardened steel (AISI 1042) were found that the cutting force was reduced by 70%. Germain et al. [8] investigated that the residual stress and surface finish produced by LAM of AISI 52,100 hardened steel, the values of surface finish (Ra) ranged between 0.6 and 0.75 mm with a feed rate of 0.1 mm/rev.

The residual stress developed to be more tensile and the hallowed penetration depth was observed when compared to the conventional cutting techniques. Dumitrescu et al. [8] showed that LAM tooth chip formation suppressed machining chatter and improved tool life by as much as 100% for AISI D2 tool steel. Most of the research reported above that focused on enhancing the functionality and machinability of LAM hard-to-machine materials, where better surface removal rates and longer tool life are typically explored in optimizing the parameters of LAM. Few studies have systemically investigated the best LAM parameter combination to achieve superior surface integrity of hardened steel components. In order to achieve the optimum surface quality of hardened steel parts, few studies have systemically examined the best LAM performance and its effectiveness.

In the present study, the laser-assisted machining of EN24 steel was studied using a traditional lathe system integrated with a 6 kW fiber delivered diode laser. A systematic laser-assisted machining study was carried out, and surface roughness and tool wear were examined by varying parameters such as laser power, spindle speed, and depth of cut. A three-dimensional transient, heat transfer model of the process was developed through the COMSOL Multiphysics software tool to compute temperature profile in the workpiece. The surface roughness and tool wear rate were compared with conventional machining process.

## 2 Experimental Work

### *2.1 Experimental Setup*

Laser-assisted machining experimental trials were carried out on the lathe machine (make: HMT). The machine has six-geared operation system with a maximum power of 2000 rpm (spindle speed) starting from 45 rpm. Spindle chucks can hold a maximum of 120 mm diameter rods. The machine is integrated with a 6 kW diode laser with the help of laser fiber through which a laser beam is pumped to the head which is fixed on the universal fixture which has five degrees of freedom. The laser beam of size 1.5 mm diameter is focused just above the tool tip with the distance between the beam and the tip of 5 and 10 mm. The tool post and the fixture were connected on the same bed so that when the tool moves, the laser head also moves at the same linear speed. Initially, EN 24 rods of 24 mm were roughly turned, as this was performed to remove any dirt or oxide layer from the surface. The workpiece was mounted inside the spindle and checked for wobble to avoid uneven machining and to avoid surface damage. The experimental setup is shown in Fig. 1.

**Fig. 1** Lathe machine is integrated with a diode laser

## 2.2 *Experimental Procedure*

Prior to machining operation, the laser spot was focused on the base material with varying power (100, 200, and 500 W) for a single rotation to understand the occurrence of heat affected zone or penetration of the heat that can be observed. At lower power, the target surface was un-melt and was also be able to soften the material up to a few microns level.

The laser head was positioned at an angle of 22.5° maintaining the working distance as the focused laser beam. An infrared thermometer was used to monitor the surface temperature of the target material before and after the LAM process. Initially, experimental trials were performed with zero laser power followed with laser power, which helps to note the comparison between the conventional and laser-assisted machining. The tool wear rate was continuously recorded before and after the machining process. The machined workpiece was further used for metallographic characterization. The tool wear and surface roughness were further analyzed with the help of a 3D Olympus DSX510 optical digital microscope.

## 2.3 *Modeling of LAM and Experimentation*

A three-dimensional transient finite element model was developed to study the temperature distributions in the workpiece, which was due to interaction with the laser beam. In the process of LAM, two different forms of heats are generated, the primary one was being the laser as a heat source, and secondly the friction generated between the tool and workpiece. The heat generated by the friction was not considered as the aim of the model to evaluate the temperatures before the interaction of the tool and the workpiece. The main objective of the developed model was to find the optimum process parameters like depth of cut and the distance to be maintained between the tool and laser beam. This further helps to reduce the wastage of the actual operation time of the machine as well as the material. COMSOL Multiphysics software package was used to model the whole process.

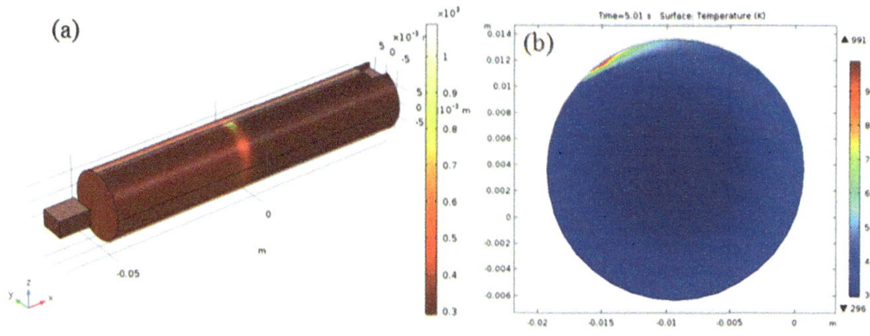


**Fig. 2** **a** Temperature profile along the surface and **b** cross section of the workpiece

The thermal distribution of the heat generated due to laser beam interaction with the target surface was analyzed in such a way that the penetration of heat into the workpiece, and the distribution on the surface area was focused. The surface temperature achieved in the experimental process, and the surface temperature of the model was also compared. By comparing these with the cross-sectional microscope imaging of the processed workpiece, the model was validated. The thermal distributions on the surface and cross section of the workpiece are shown in Fig. 2.

Figure 2b describes the temperature at the tail of the heated region which is approximately 500–600 k. It was observed that at this temperature, the material is soft enough to be easily machined. The exact location, where the tool tip, should come in contact with the workpiece. Similarly, the heat penetration in the cross section till approximately 400 microns was noticed as a reaching temperature, where the target material was soft enough to be cut easily. This was decided and proven to be a very efficient depth of cut for the practical operation.

## 3 Results and Discussions

The turning operation was carried out with the reference of simulation studies. The rough turning operation was done on EN 24 steel rods. The laser mounted lathe machine controls and gears were checked in proper order before starting the experiment. The workpiece was set inside the spindle and checked for wobble; after this step, it was cleaned for processing. Initially, the samples were taken with only laser-focused on the workpiece in a single rotation, so, therefore, a couple of 360° laser-focused experiments were carried out at different powers of 100 and 500 W. Several experimental trials were carried out to optimize the process parameters. Finally, the laser power of 100 W was maintained constant with varying spindle speed, depth of cut, and feed rate. Table 1 depicts the process parameters and the obtained surface roughness values of the LAM process of EN24 steel.

**Table 1** Optimized parameters of the LAM process of EN24 steel

| S. No. | Depth of cut (mm) | Spindle speed (rpm) | Feed rate (mm/rev) | Surface roughness |
|---|---|---|---|---|
| 1 | 0.2 | 1000 | 0.6 | 1.458 |
| 2 | 0.2 | 1600 | 0.375 | 0.994 |
| 3 | 0.2 | 1000 | 0.6 | 0.5 |
| 4 | 0.2 | 1600 | 0.375 | 0.493 |
| 5 | 0.4 | 1000 | 0.6 | 0.869 |
| 6 | 0.4 | 1600 | 0.375 | 0.927 |
| 7 | 0.4 | 1000 | 0.6 | 1.204 |
| 8 | 0.4 | 1600 | 0.375 | 1.1 |

It was observed that while LAM experimental process, the surface roughness was noticed to be reduced by increasing the temperature of material removal with laser heating. The decrease in surface roughness is a change in depth of cut, when compared to that of the conventional machining as shown in Fig. 3a. It also can be observed that the surface of the machined surface of LAM is better than the conventional machined surface as shown in Fig. 3b.

Figure 4a shows a tool used for conventional machining, where more tool wear can be observed and similarly which on comparing to Fig. 4b shown was used for the LAM process, and it has more worn in comparison with the conventional machining. It was observed that the reason for more wear was due to tool melting, which was mainly due to the heat produced by the laser and the focused distance. In this case, the laser power was more (200 W), so more optimization was done based on analysis was carried out. The distance between the tool and the beam was increased to 10 mm,

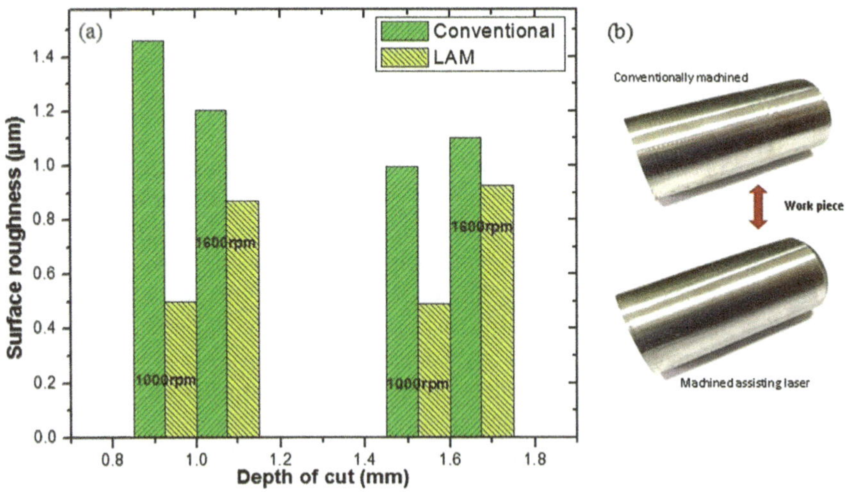


**Fig. 3** **a** Effect of depth of cut on surface roughness, **b** machined surfaces

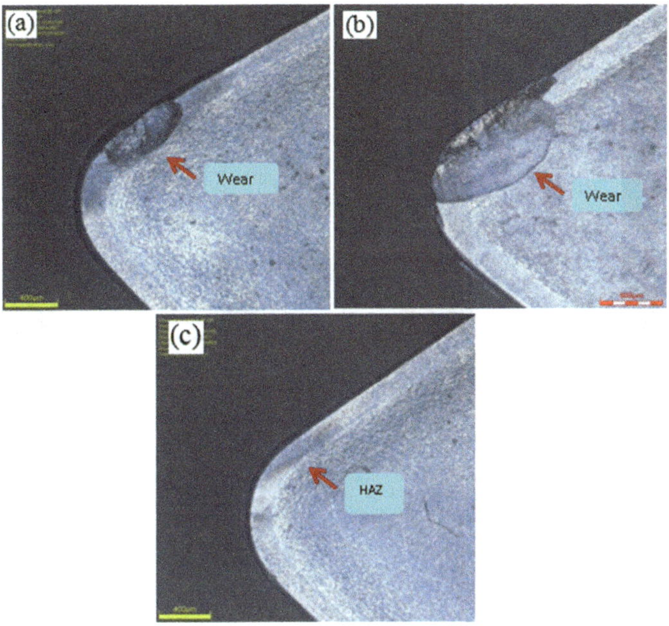


**Fig. 4** Tool used for **a** Conventional machining, **b** LAM with 200 W laser power, **c** LAM with 100 W power

and laser power was decreased to 100 W. In Fig. 4c, tool wear can be noticed almost nil compared to conventional machining, which was only due to LAM with 100 W power and increased distance between the tool tip and laser beam, heat-affected zone was found on the material removal tip.

It can be observed from Table 1 that the optimized parameters showed better results when compared to the conventional methods. All the processed target materials showed better results on decreasing the power and increase the gap between the beam and tool tip. The tool wear plots are shown in Fig. 4, where tool weight was measured after each run of the process, and the difference in weight was recorded and plotted in the graphs, which shows that tool wear in conventional method is more when compared to laser-assisted machining. The spindle speeds were considered as 1000 rpm and 1600 rpm, whereas the depth of cut was maintained as 0.2 mm, 0.4 mm and compared with each other is shown in Fig. 5a and b. Hence, it can be said that the laser-assisted machining has improved to resistance to tool wear when compared to the conventional method. The above study proves that the laser-assisted machining has advantages like better finishing, improved tool wear, and economical at an industrial scale.

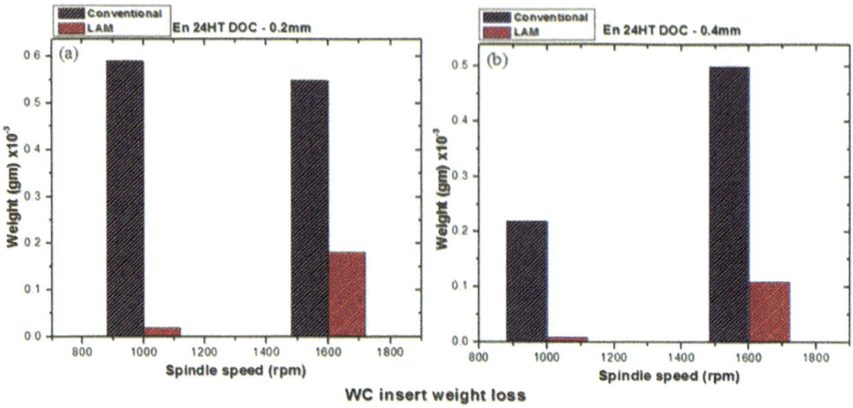


**Fig. 5** Tool wear rate of conventional versus LAM process of **a** 0.2 mm depth of cut, **b** 0.4 mm depth of cut

## 4 Conclusions

An experimental study of LAM on EN24 steel was successfully carried out under several process parameters using WC tool. The LAM performance was compared to that of the conventional machining. The following conclusions are summarized as follows.

- In laser-assisted machining, heat penetration (heat affected zone) and surface roughness were controlled by various operating parameters.
- The laser power of 100 W was maintained constant with varying spindle speed, depth of cut, and feed rate.
- The surface roughness was decreased by an increase in the temperature of the target material during laser heating. This was observed as a change in depth of cut when compared to that of the conventional machining.
- The tool wear was almost nil with lower power of LAM as compared to conventional machining, heat-affected zone was found on the material removal tip.
- Tool weight was measured after each run of process, and the difference in weight was recorded. The tool wear of the conventional method was more when compared to laser-assisted machining. This proves laser-assisted machining has improved the resistance to tool wear when compared to the conventional method.

## References

1. Liang S (2006) Technology assessment on current advanced research projects in the area of hard turning. The Association of Manufacturing Technology

2. Konig W, Zaboklicki AK (1993) Laser-assisted hot machining of ceramics and composite materials. In: International conference on machining of advanced materials, vol 847. NIST Special Publication, Gaithersburg, MD, pp 455–463
3. Courtney C, Steen WM (1979) Laser surface treatment of EN8 steel using a 2 kW $CO_2$ laser. Met Technol 6(12):456–462
4. Rozzi JC, Shin YC, Incropera FP (2000) Experimental evaluation of the laser assisted machining of silicon nitride ceramics. Journal of Manufacturing Science and Engineering, Transactions of the ASME 122:666–670
5. Lei S, Shin YC, Incropera FP (2001) Experimental investigation of thermo-mechanical characteristics in laser assisted machining of silicon nitride ceramics. J Manuf Sci Eng Trans ASME 123:639–646
6. Gratias JF, Fan LJ, Marot G, Cohen P (1993) Proposition of a method to optimize the machining of XC42 steel with laser assistance. CIRP Ann 42:115–118
7. Germain G, Morel F, Lebrun JL (2006) Effect of laser assistance machining on residual stress and fatigue strength for a bearing steel ($100Cr_6$) and a titanium alloy (Ti6Al4V). In: Materials science forum, residual stresses VII—Proceedings of the seventh European conference on residual stresses, ECRS7, pp 524–525, 569–574
8. Dumitrescu P, Koshy P, Stenekes J, Elbestawi MA (2006) High-power diode laser assisted hard turning of AISI D2 tool steel. Int J Mach Tools Manuf 46:2009–2016

# Optimization of Wire Cut Electric Discharge Machining Characteristics of Hybrid Aluminium Composites (Al6061/Gr/SiCp) Using Taguchi Method

**P. Gavisiddesha, C. Thotappa, Veerabhadrappa Algur, and B. Suresh Reddy**

**Abstract** This paper presents investigation and optimization of machining characterization of hybrid aluminium 6061 reinforced with graphite and 4% weight percentage of SiC using Taguchi approach. The four process parameters chosen were pulse-on time, pulse-off time, servo voltage (V) and wire speed, and output parameters were material removal rate and surface roughness. The experiments were conducted based on Taguchi L9 (3^4) orthogonal array. The results obtained from this investigation presumed to be applicable for to choose appropriate wire EDM process parameters to machine hybrid Al6061 reinforced with graphite and silicon carbide particles with 4% (weight percentage).

**Keywords** Orthogonal array · Stir casting process · MRR · Surface roughness · Taguchi · Wire electric discharge machining

## 1 Introduction

Nowadays, most of the applications like marine, military and structural equipment's selection of materials are done according to their properties like high strength-to-weight ratio, high durability and so on. It is difficult for monolithic alloy to have all the required properties; therefore, composite materials came into picture [1]. Many research studies are done on various materials to find their properties to suit their

P. Gavisiddesha (✉)
Ballari Institute of Technology and Management, Ballari, Karnataka, India

C. Thotappa · V. Algur
Department of Mechanical Engineering, Rao Bahadur Y Mahabaleswarappa Engineering College, Ballari, Karnataka, India

B. Suresh Reddy
Rao Bahadur Y Mahabaleswarappa Engineering College, Ballari, Karnataka, India



P. Srinivasa Pai and V. Krishnaraj (eds.), *Sustainable Machining Strategies for Better Performance*, Lecture Notes in Mechanical Engineering,
https://doi.org/10.1007/978-981-16-2278-6_5

applications in engineering industries. Among many of the materials, composite materials possess excellent properties, and hence, they find its uses in various industries [2–5]. Sometimes adding a single reinforcement material to an alloy leads to surplus its properties, and hence, more than one (hybrid) materials will be added to overcome the drawback of single metal matrix composites. Thus, hybrid metal matrix [HMMC] concept came into existence. Silicon carbide consists of two important elements, namely silicon (Si) and carbon (C). Moissanite is the rare mineral in which silicon carbide is present in crystalline form. The physical property of pure silicon carbide is usually transparent and colourless crystal. When an impurity such as aluminium is added, it changes into greenish or light blue colour based on percentage of impurity added. Silicon carbide consists of four carbon atoms, covalently bonded to a single silicon atom at centre. The main property of silicon carbide is high refractoriness; i.e. it possesses high melting temperature and low thermal expansion, and hence, it is used as refractory lining material in high-temperature applications [6–10]. In addition to above properties, it is hardest material next to boron carbide and diamond, and this apparent property makes silicon carbide to be used in most of the cutting tools as it has high wear resistance. However, it has superior properties compared to aluminium like high strength, high melting point, and stiffness and thus enhances the basic qualities of aluminium when combined [11–15]. Moreover, combined ceramic and semiconductor properties of silicon carbide find its application in manufacturing of high-voltage and high-temperature devices. On the other hand, increase in hardness of silicon carbide leads to increase in cutting tool wear rate which has negative effect on machinability of Al alloy [16]. Different machining operations like milling, turning and drilling of Al/SiC composite use hard nitride coated tools and diamond to produce desired shaped composites which are being used in aeronautical applications. To overcome this limitation, one more reinforcement material is added to improve machinability of Al alloy. In order to slow down the wear rate of cutting tool, proper lubricating material should be added to withdraw the heat and to enhance tool life. Graphite is a mineral with many thin layers of carbon atoms. Graphite is formed by a process called as metamorphism in which reduction of sedimentary carbon compounds takes place leads to formation of graphite. It is a natural mineral and extensively used in manufacturing of pencils. Graphite consists of hexagonal crystalline structure forming thin layers of graphenes. Compared to diamond, the carbon atom bonding is found to be vigorously stronger in graphite. Many research studies are carried out on graphite properties and their applications and outcome of those studies resulted with higher electrical and thermal conductivity and can work in high-temperature applications, the most predominant property of graphite is self-lubricating and is corrosive resistance to some chemicals and acids. Undoubtedly, Al 6000 series possesses well-defined mechanical characteristics compared to its other alloys. Similarly, silicon carbide also shows better mechanical characteristics. Hence for further improving the properties of Al alloy, silicon carbide is added to enhance properties like strength and hardness in the Al alloy. The vital zeal of the work is to prepare aluminium hybrid composite [HMMC]. The raw materials are processed by stir casting and are prepared as per ASTM standards. These samples are subjected to evaluation of machinability in wire cut EDM process. Anthony Xavior [17] studied on machinability of hybrid composites and effect of feed rate, depth of cut, cutting speed on cutting force, surface roughness of hybrid MMC and concluded that surface roughness mainly depends on feed rate; i.e. lower feed rate will have lesser surface roughness. The presence of graphite in MMC improves surface finish. On the other hand, graphite-based MMC requires more cutting force. Nagendra Kumar Mayor [18] studied mechanical characteristics of aluminium 6061 alloy reinforced with silicon carbide and focussed on distribution of silicon carbide in Al alloy. Based on the results, he concluded that density of Al-silicon carbide composite is increased but the presence of silicon carbide in Al alloy improves both tensile strength and hardness compared to Al 6061 alloy alone. Palani Kumar et al. [19] considered the effect of drilling parameter on thrust force in drilling Al 661/15% SiC 4%Gr metal matrix composites. The results revealed that feed rate on thrust force is the key parameter which influences in drilling of hybrid metal matrix composites. The increase in feed rate also increases the burr formation. Based on his results, it is clear that at higher spindle speeds, thrust force reduces, further improves surface finish. He concluded that lesser feed rate, higher spindle speed values and smaller diameter of drill bit improve surface finish in a hybrid MMC. The presence of graphite in MMC reduces the hardness of composite which is favourable for machinability of composites. Sivananthan [20] studied on aluminium–silicon carbide composite, tensile and hardness values, and results of his experiments reveal that hardness, tensile property and compression strength of Al–silicon carbide composite have got improved in their properties by diffusing increased weight fraction of silicon carbide reinforcement material with Al6061 alloy [21–25]. Hence, Al–Silicon carbide composites are widely used as good composites especially in nuclear and missile industry. The present study reveals that the investigation and optimization of machining characterization of hybrid aluminium 6061 reinforced with graphite and 4% weight percentage of SiC particles using Taguchi approach.

## 2 Experimental Methodologies

### *2.1 Work Sample Preparation*

The furnace consists of heating coils which are wounded around crucible placed at the centre. Electrically driven stirrer with impeller is placed inside the crucible during melting of raw materials. Before starting the process, the crucible is cleaned and is placed at furnace centre. Al6061 ingots of required weight are weighed in weighing balance, and then it is dumped into crucible. Switch-on the furnace and set the working temperature to 750-800 °C. During melting of Al6061, required weight fractions of reinforcement materials (4% Wt of SiC and 3% Wt of Gr) and then place it in crucible. Electrically driven stirrer coated with ceramic or graphite is then operated inside crucible for uniform distribution of reinforcement materials with base metal. Once molten liquid attains pouring temperature, crucible is removed

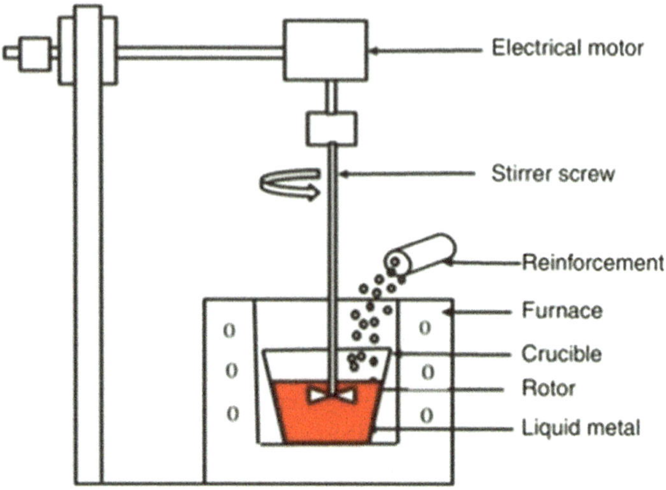


**Fig. 1** Stir casting set-up

from the furnace and molten hot liquid is poured into pre-heated metallic moulds of required dimensions. After solidification, the cast is removed from mould box and it is cool down to room temperature. Stir casting set-up is shown in Fig. 1.

## 2.2 *Machine Tool*

The machining operations were performed on Sodick wire electric discharge machine set-up at Navanith Enterprises, Bangalore. EDM process uses high-frequency DC power. During machining, wire acts as anode and workpiece acts as cathode. Pulse current value was 17 A, and pulse voltage was 50V. The dielectric fluid usually de-ionized water is used to carry electric charge/sparks in the air gap between two electrodes. The dielectric fluid can be reused for each cycle through proper filtering system The brass wire of 0.25 mm diameter is continuously feed through wire guide wheel under given tension and after machining it is not wounded in a take up wheel since it is consumable electrode. When power supply is ON, the wire (electrode) starts emitting electric sparks in the gap between wire and work sample. The dielectric fluid carries energy of sparks towards work surface and machining takes place by melting and vapourizing the surface. It also carries away the machined debris from machined surface and provides clean surface for next machining cycle. Machine used to perform machining is shown in Fig. 2.

This study involves machining of HMMC's using wire EDM by considering four input variable factors, namely pulse-on, pulse-off time, voltage and wire speed. Material removal rate and surface roughness are needed to be calculated under three different levels of input variables. Dr. Genichi Taguchi is one of a famous total quality

**Fig. 2** Sodick wire EDM set-up

management guru who introduced Taguchi method to find out good set of experimental values. It involves both statistical and mathematical models and is widely been using by many engineers and economists. It provides a systematic way to carry out experiments and to find best or optimal experimental values [26, 27]. Surface roughness of each machined component is determined by using surface roughness tester 'mitutoyo talysurf' as shown in Fig. 3.

Similarly, another output response, i.e. metal removal rate, is evaluated as the ratio of difference of weight of test sample before and after wire EDM machining with respect to time required for cutting (machining time).

$$\text{MRR} = (\text{Wb - Wa})/(T) \quad \text{g/min}$$

where

Wb = Weight of test sample before machining.

Wa = Weight of test sample after *m*.

$T$ = Machining time.

The output responses like MRR and surface roughness obtained from the experimental values are then transferred into signal-to-noise ratio (S/N). This S/N ratio helps to optimize the parameters and serves as a better tool in optimum result prediction. Signal-to-noise ratio can be analysed by three quality levels such as larger the better type (LTBT); nominal is the better type (NTBT) and smaller the better type (STBT). Below shown expressions are used to check three quality characteristics. S/N ratio is tabulated by using above shown logarithmic functions.

$$\text{Smaller is better:} \quad \frac{S}{N} = 10 \log \frac{1}{n}\left(\sum y^2\right)$$

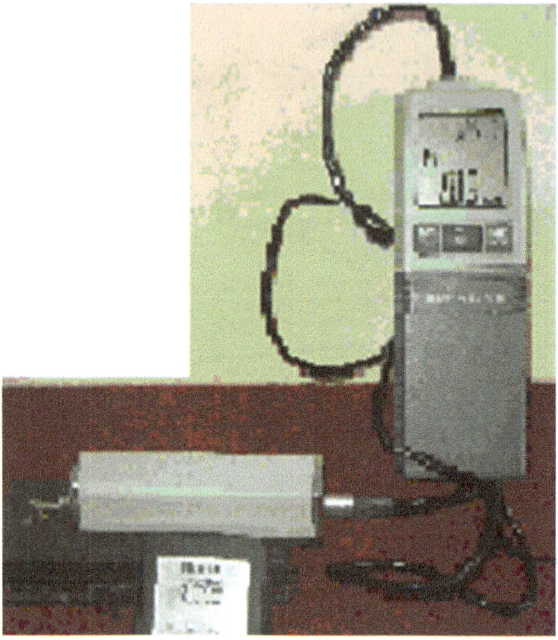

**Fig. 3** Surface roughness measuring instrument

$$\text{Larger is better:}\quad \frac{S}{N} = 10\log\left(\sum \frac{\ddot{Y}}{S^2}\right)$$

$$\text{Nominal is best:}\quad 10\log\frac{1}{n}\left(\sum \frac{1}{y^2}\right)$$

where

'$n$' represents total number of trials,
'$y$' indicates output value or observed data,
$\hat{Y}$ is the mean value,
'$S$' is the variance.

## 3 Results and Discussions

This chapter explains the effect of each four control factors on two response outputs of HMMC's considered under this study. After tabulating the MRR and Surface roughness values, the machining parameters need to be optimized in order to gain better output responses. Taguchi developed a technique, which was built on traditional concepts of design of experiment (DOE). Orthogonal array (OA) is a specially constructed table used for the experimentation purpose, based on DOE technique to reduce the number of experiments, so as to give the optimum set of the preferred parameters. First step in Taguchi is to create orthogonal array by considering number of input and output responses. Later the obtained results are analysed with S/N ratio to optimize process parameters [28]. Experiment is carried out by considering four input variables and two output responses at three levels.

For optimizing the machining parameters, Minitab-17 version software is used. This software helps to find optimized machining parameters values by calculating signal-to-noise ratio for each of the trials. Based on Taguchi L9 orthogonal array, nine trials are carried out on test sample C.

Later by using equation of metal removal rate is calculated using digital weighing balance with precision of 1 mg and with the help of Mitutoyo surface tester, surface roughness of nine machined sub-samples is tabulated. The obtained values are then converted into signal-to-noise ratio as shown in Table 1. To find the effect of control factors on MRR, higher the better S/N ratio values are considered and for surface roughness, smaller the better values are considered. Response table provides information regarding which control factor will have major effect on outputs followed by least (minor) effect. This can be noted by referring rank row in each of the response table. The L9 orthogonal array was chosen to conduct the experiment. Test sample C is machined through wire EDM to study machinability, and nine machined samples named from C1 to C9 are shown in Fig. 4.

Figure 5 implies metal removal rate that increases along with increase in control factor, namely pulse-on time. Under high spark energy, material gets more heated up resulting in high MRR. But increase in servo voltage is inversely varying with MRR, because at higher servo voltage more heat energy is wasted without heating work piece and hence lesser MRR. While MRR initially increases at higher values of wire speed factors and then decreases. Moreover, pulse-off time has negligible effect for MRR. By referring Table 2, MRR mainly depends on pulse-on time and servo voltage followed by wire speed and pulse-off time. Therefore to get higher MRR values, the process parameters need to be optimized. The optimized values are A3 B1 C1 D2.

Similarly, for surface roughness, the main effect plots are plotted as shown in figure along with response table. But for optimization, S/N ratio is considered under smaller is better type. By referring Fig. 6 and Table 3, surface roughness is reversely related with pulse-on time variable and wire speed, whereas it increases with increase in servo voltage and pulse-off time. Thus for smaller surface roughness responses, the optimized parameters are A1B1C3D1.

## 4 Conclusion

Al6061 reinforced with SiC and Gr HMMC's is fabricated by stir casting technique and is machined successfully by using wire electric discharge machining process. The experiment uses Taguchi method to optimize wire electric discharge machining characteristics. Machining performance characteristics evaluated are material removal rate and surface roughness. The four process parameters chosen were pulse-on time,

**Table 1** Experimental design using L9 orthogonal array for test sample C

| S. No. | Pulse-on (μs) | Pulse-off (μs) | Servo voltage (V) | Wire Speed (m/min) | Machining time in (min) | MRR (gm/min) | Surface roughness ($R_a$) (μm) | S/N of MRR | S/N of Ra |
|---|---|---|---|---|---|---|---|---|---|
| 1 | 2 | 15 | 30 | 1 | 11.52 | 0.868 | 3.60 | −1.22961 | −11.1261 |
| 2 | 2 | 16 | 40 | 2 | 12.52 | 0.878 | 3.64 | −1.13011 | −11.2220 |
| 3 | 2 | 17 | 50 | 3 | 14.34 | 0.697 | 3.60 | −3.13534 | −11.126 |
| 4 | 4 | 15 | 40 | 3 | 9.3 | 1.180 | 4.00 | 1.43764 | −12.0412 |
| 5 | 4 | 16 | 50 | 1 | 11.5 | 0.869 | 4.03 | −1.21960 | −12.1061 |
| 6 | 4 | 17 | 30 | 2 | 8.2 | 1.341 | 4.10 | 2.54858 | −12.2557 |
| 7 | 6 | 15 | 50 | 2 | 8.1 | 1.234 | 4.00 | 1.82630 | −12.0412 |
| 8 | 6 | 16 | 30 | 3 | 7.5 | 1.333 | 4.33 | 2.49660 | −12.7298 |
| 9 | 6 | 17 | 40 | 1 | 8 | 1.250 | 3.99 | 1.93820 | −12.0195 |

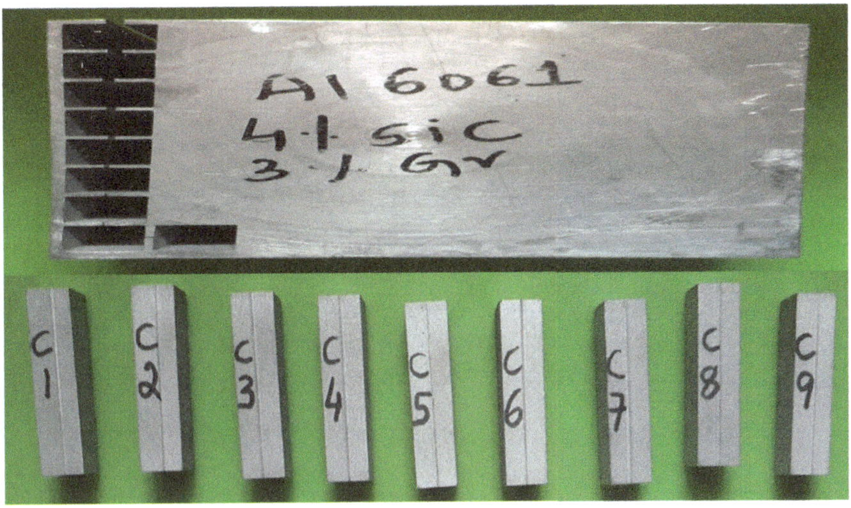


**Fig. 4** Sample C and machined pieces of sample C

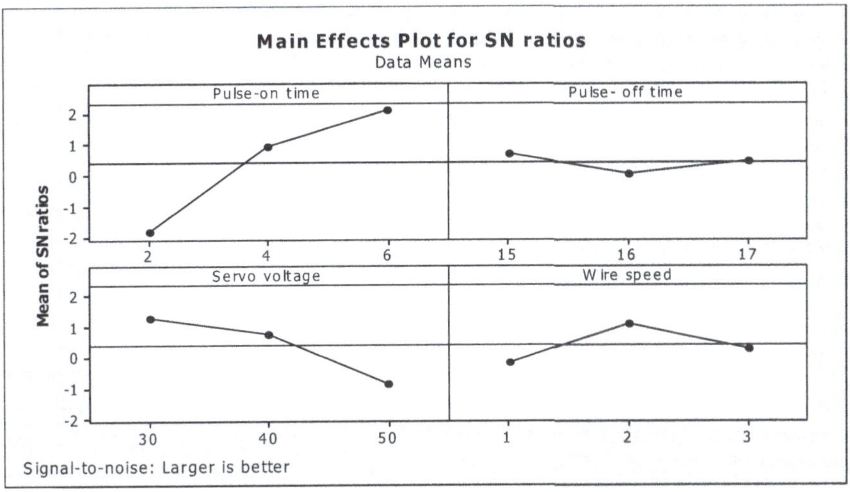


**Fig. 5** Main effects plot for SN ratios of MRR for sample C

**Table 2** Response table for signal-to-noise ratios of MRR for sample C (larger is better)

| Level | Pulse-on time | Pulse-off time | Servo voltage | Wire speed |
|---|---|---|---|---|
| 1 | −1.83169 | 0.67811 | 1.27186 | −0.17034 |
| 2 | 0.92220 | 0.04896 | 0.74858 | 1.08159 |
| 3 | 2.08704 | 0.45048 | −0.84288 | 0.26630 |
| Delta | 3.91872 | 0.62915 | 2.11474 | 1.25193 |
| Rank | 1 | 4 | 2 | 3 |

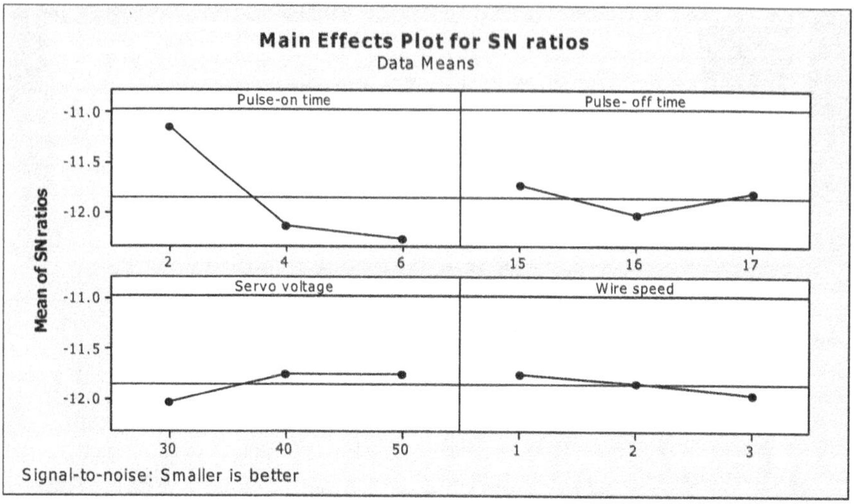


**Fig. 6** Main effects plot for SN ratios of surface roughness for sample C

**Table 3** Response table for signal-to-noise ratios of surface roughness for sample C (smaller is better)

| Level | Pulse-on time | Pulse-off time | Servo voltage | Wire speed |
|---|---|---|---|---|
| 1 | −11.16 | −11.74 | −12.04 | −11.75 |
| 2 | −12.13 | −12.02 | −11.76 | −11.84 |
| 3 | −12.26 | −11.80 | −11.76 | −11.97 |
| Delta | 1.11 | 0.28 | 0.28 | 0.22 |
| Rank | 1 | 2 | 3 | 4 |

pulse-off time, servo voltage and wire speed, and output parameters were material removal rate and surface roughness. In the initial stage of experiment, evaluation of effect of process parameters which mainly affect the output parameters was studied. For this study, the range of process parameters also selected nine steps is performed for this experiment by using Minitab-17 software. According to the experimental studies, conclusions are made as follows.

- Using Taguchi method, metal removal rate (MRR) and surface roughness (Ra) were optimized for the prepared test sample, and the optimal values of process parameters on metal removal are pulse-on time-6 (micro-sec), pulse-off-15(micro-sec), servo voltage-30(V), wire speed-2(m/min).
- Optimal values of process parameters for surface roughness are pulse-on-2 (micro-sec), pulse-off-15 (micro-sec), servo voltage-50 (V), wire speed-1 m/min.
- The most significant process parameter that affects metal removal rate (MRR) is pulse-on time while pulse-off has no effect.

- For surface roughness, most potential process parameter is pulse-on time while wire speed has no effect.

## References

1. Singh G, Singh MP, Singh G (2013) Optimization of the machining parameters for surface roughness during turning of Al/SiC/Gr Hybrid MMC. Int J Eng Res Tech 2(11):1613–1617
2. Sasimurugan T, Palanikumar K (2011) Analysis of the machining characteristics on surface roughness of a hybrid aluminium metal matrix composite (Al6061-SiC-$Al_2O_3$). J Miner Mater Charact Eng 10(13):1213–1224
3. Aruri D, Adepua K (2013) Ŵear and mechanical properties of 6061–T6 aluminum alloy surface hybrid composites [(SiC+ Gr) and (SiC+ Al2O3)] fabricated by friction stir processing. J Mater Res Technol 2(4):362–369
4. Manna A, Bhattacharyya B (2003) A study on machinability of Al/Sic-MMC. J Mater Process Technol 140:711–716
5. Lokesh T, Mallikarjun US (2014) Mechanical and morphological studies of Al6061-Gr-SiC hybrid metal matrix composites. IJRMET 4(2):135–138
6. Suresh P, Marimuthu K, Ranganathan S, Rajmohan T (2014) Optimization of machining parameters in turning of Al−SiC−Gr hybrid metal matrix composites using grey-fuzzy algorithm, Elsevier. Trans Nonferrous Met Soc China 24:2805–2814
7. El-Gallab M, Sklad M (1998) Machining of Al/SiC particulate metal-matrix composites part I: tool performance. J Mater Process Technol 83:151–158
8. Krishna MV, Xavior AM (2014) An investigation on the mechanical properties of hybrid metal matrix composite. In: 12th Global congress on manufacturing and management. Elsevier, pp 918–924
9. Altunpak Y, Mustafa A, Aslan S (2012) Drilling of a hybrid Al/SiC/Gr hybrid metal matrix composites. Int J Adv Manuf Technol 513–517
10. Kathiresan M, Sornakumar T (2010) EDM studies on aluminum alloy-silicon carbide composites developed by vortex technique and pressure die casting. JMMCE 9(1):79–88
11. Velmurugan C, Subramanian R, Thirugnanam S, Ananadavel B (2011) Experimental investigations on machining characteristics of Al 6061 hybrid metal matrix composites processed by electrical discharge machining. Int J Eng Sci Technol 3(8):87–101
12. Kumar GBV, Rao CSP (2012) Selvaraj N (2012) Studies on mechanical and dry sliding wear of Al6061–SiC composites. Compos B 43:1185–1191
13. Shandilya P, Jain PK, Jain NK (2011) Wire electric discharge machining of metal matrix composite materials. Danube Adria Association for Automation & Manufacturing, Scientific Book, vol 10, pp 383–400
14. Muniappan A, Anandan D, Aravind D, Dhanraj K, Abinesh R, Barathkumar P (2015) Performance of wire electrodes on machining of hybrid Aluminium composite. IJAER 10(65)
15. Garg KH, Kumar R, Manna A, Sharma S (2019) Mathematical modelling and parametric optimization of EDM for tool wear rate of hybrid aluminum metal matrix composite reinforced with SiCp and Grp. Int J Emerg 10(2):307–316
16. Akshay D, Pradeep K, Inderdeep S (2008) Experimental investigation and optimization in EDM of Al 6063 SiCp metal matrix composite. Int J Mach Mach Mater 5:293–308
17. Anthony Xavior M, Ajith Kumar JP (2017) Machinability of hybrid metal matrix composite - a review. Proceedia Eng 1110–1118
18. Maurya NK, Maurya M (2020) Investigation of mechanical properties of Al 6061/SiC composite prepared through stir casting technique. Mater Today Proc 25, 755–758
19. Palanikumar K, Muniaraj A (2014) Experimental investigation and analysis of thrust force in drilling cast hybrid metal matrix (Al–15%SiC–4%graphite) composites. Measurement 240–250

20. Sivanathan S, Ravi K (2020) Effect of SiC particles reinforcement on mechanical properties of aluminium 6061 alloy processed using stir casting route. Mater Today Proc 21:968–970
21. Nagarala M, Shivananda BK (2017) Mechanical behavior of Al6061-Al2O3 and Al6061-graphite composites. Mater Today: Proc 10978–10986
22. Gopalsamy BM, Mondal B, Ghosh S (2009) Taguchi method and ANOVA: An approach for process parameters optimization of hard machining while machining hardened steel. J SCI IND RES 68:686–695
23. Muthukrishnan N, Davim JP (2009) Optimization of machining parameters of Al/SiC-MMC with ANOVA and ANN Analysis. J. Mater. Process 209:225–232
24. Riaz Ahamed A, Asokan P, Aravindan S (2009) EDM of hybrid Al–SiCp–B4Cp and Al–SiCp–Glassp MMCs. Int J Adv Manuf Technol 520–528
25. Radhika N, Sudhamshu AR (2014) Optimization of electrical discharge machining parameters of aluminium hybrid composites using Taguchi method. JESTEC 9(4):502–512
26. Algur V, Kabadi VR, Ganechari SM, Shetty PB, Hulipalled P (2014) Analysis of wear behaviour of a heat treated modified ZA-27 alloy by Taguchi technique. Int J Recent Innov Trends 2(12)
27. Algur V, Jayprakash KG, Praveen, Unki S, Diggavi S, Experimental investigation on dry sliding wear behavior of Plasma sprayed $TIO_2$–Inconel 718 coatings on A16061 by using Taguchi Technique. JETIR 6(5):181–185. ISSN: 2349-5162
28. Chauhan VR, Dinesh KR, Veeresh K, Algur V, Shettar M (2017) Effect of hematite ore on wear behaviour of e-glass fibre reinforced with polyster resin-a Taguchi approach. IJERMT 06(05). 10.23956. ISSN: 2278-9359

# Minimum Quantity Lubrication and Cryogenic for Burnishing of Difficult to Cut Material as a Sustainable Alternative

**B. Sachin, Charitha M. Rao, Gajanan M. Naik, C. Durga Prasad, Ajit M. Hebbale, V. Vijeesh, and Muralidhara Rao**

**Abstract** To reduce the harmful effects of the lubricants, adequate strategies are demanded to incorporate sustainable manufacturing techniques. Cryogenic cooling is one of the effective lubricants which can improve the surface characteristics of the material. Generally, machining under various lubrication systems is performed on difficult-to-cut materials. The material considered is 17-4 precipitation-hardenable (PH) stainless steel which is known for its corrosion resistance properties. Diamond burnishing is one of the superfinishing processes which is used to improve the surface integrity characteristics of the materials. The primary focus of this work is to investigate the impact of diamond burnishing on surface morphology, subsurface microhardness, and residual stress of 17-4 PH stainless steel in sustainable cooling environments. Enhancement in the performance characteristics was noticed under the cryogenic environment. It was perceived that the cryogenic environment is the better mode of lubrication in comparison with the other modes of cooling/lubrication.

**Keywords** Sustainable cooling environments · Residual stress · Subsurface hardness · Morphology

---

B. Sachin (✉)
Department of Mechanical Engineering, The National Institute of Engineering, Mysuru, Mysuru 570008, India

C. M. Rao
Department of Mechanical Engineering, Faculty of Engineering and Technology, Jain (Deemed-to-be) University, Bengaluru 562112, India

G. M. Naik
Department of Mechanical Engineering, Mangalore Institute of Technology and Engineering, Moodabidre, Mangalore, Karnataka 574225, India

C. Durga Prasad
Department of Mechanical Engineering, RV Institute of Technology and Management, Bengaluru 560076, India

A. M. Hebbale · V. Vijeesh · M. Rao
Department of Mechanical Engineering, NMAM Institute of Technology, Nitte 574110, India

 
P. Srinivasa Pai and V. Krishnaraj (eds.), *Sustainable Machining Strategies for Better Performance*, Lecture Notes in Mechanical Engineering,
https://doi.org/10.1007/978-981-16-2278-6_6

# 1 Introduction

Surface modification is one of the essential methods used in the present scenario to improve the performance of the product under varying loading conditions. Most of the manufacturing industries will opt for secondary finishing processes to perform the surface modification. In the conventional methods, the surface defects encountered during the machining on the surface of the material is unavoidable. Hence, there is a need to explore the most suitable technique to minimize the problems caused by conventional machining. Burnishing is a chipless secondary finishing process performed on a machined surface to achieve the super finish of the components [1]. The material used in the research has the following characteristics: low thermal conductivity and good corrosion resistance [2–7].

Tobola et al. [8] performed gas nitriding and slide burnishing on Ti–6Al–4V ELI alloy. The surface hardness of the material was improved by 5–10% with the application of this process. Swirad [9] analyzed the surface texture of cylindrical components by using slide burnishing. Korzynska et al. [10] explored a new technique microsphere flow burnishing to plastically deform the material. The surface texture of 2017A alloy has been improved after the implementation of the new method. Banh et al. [11] developed a novel double-spring mechanism burnishing tool and conducted experiments on STAVAX material. It was found that the surface finish and superficial hardness were enhanced by 91% and 8%, respectively, using grease as the lubricant.

One of the most environmentally benign and beneficial lubrication methods which have been used in most of the manufacturing industry in improving the performance of the component is cryogenic cooling. It has a unique property of precisely engineer the surface and subsurface properties of the material. In most of the applications where difficult to cut materials are routinely used to manufacture products, cryogenic cooling is the most relevant technology to attain improved performance of the product since it is a sustainable and societally beneficial method. The use of the liquid nitrogen ($LN_2$) at the working zone can significantly improve the surface integrity of the material [12, 13]. It has been recommended by most of the researchers to achieve improved functional performance over MQL and dry environments [14–21]. The corrosion resistance and surface finish of the titanium alloy were enhanced by the use of cryogenic burnishing [22]. Huang et al. [23] analyzed the effects of cryogenic burnishing on Al 7050-T7451 alloy and observed refined nano-grains after cryogenic burnishing when compared to dry burnishing with increased hardness. Improvement in the surface integrity of Co-Cr–Mo biomaterial was observed after performing cryogenic burnishing [24].

In this article, slide burnishing was performed to explore the possibilities of improving the surface morphology, residual stress, and subsurface hardness of the material under sustainable cooling environments.

**Table 1** Conditions used for the experiments

| Process parameters | Unit | Levels | | | | |
|---|---|---|---|---|---|---|
| | | 1 | 2 | 3 | 4 | 5 |
| Burnishing speed (*S*) | m/min | 21 | 30 | 47 | 73 | 113 |
| Burnishing feed (*f*) | mm/rev | 0.048 | 0.055 | 0.065 | 0.079 | 0.096 |
| Burnishing force (*F*) | N | 20 | 50 | 90 | 120 | 150 |

## 2 Experimental Method

'Kirloskar' lathe was used to burnish the 17-4 PH stainless steel. The burnishing process was carried out under different environments, and the experimental setup can be found in [7]. Table 1 represents the details of the experimental conditions. SJ301 Mitutoyo Surftest was used to measure the surface finish. The surface roughness tester consists of a stylus tip of a radius 2 μm. A tank of capacity 55 liters was used to store the $LN_2$, and it was pressurized by sending the compressed air to the tank. SEM and microhardness tester used in the present study are 'JEOJSM-638OLA' and OMNI TECHMVHS-AUTO, respectively. MQL setup of model 'DAOML-2/PS/FS/1' was used to spray the mist of oil. Residual stress has been measured by X-ray diffraction make 'PROTO.'

## 3 Results and Discussion

### *3.1 Morphology*

The images shown in Fig. 1a–c were captured at the conditions of burnishing force = 90 N, burnishing feed = 0.055 mm/rev and burnishing speed = 47 m/min. At the cryogenic condition, the surface finish noticed was 0.10 μm. It is because in this environment the cooling of the workpiece surface with the help of $LN_2$ reduces the temperature and produces a uniform surface. Under the cryogenic environment, the metal which is accumulated at the top surface layer of the material starts flowing owing to the constant pressure of the $LN_2$ supplied at the burnishing zone. This causes the easy flow of the material. Some of the microvoids present over the surface will be filled because of the flow of the metal, and also, while flowing over the surface layer of the material, the metal occupies the space on the microvoids instantly. Hence, uniform surface will be generated because of the combination of the diamond burnishing process and cryogenic cooling effect. The feed marks have been generated during the burnishing under dry and MQL conditions. The surface finish of 0.18 μm was observed under MQL, and it was noticed to be 0.24 μm in the dry condition. Also, the presence of microvoids has been observed in the MQL and dry environments. It is due to the fact that in the MQL environment, at the applied levels of diamond burnishing

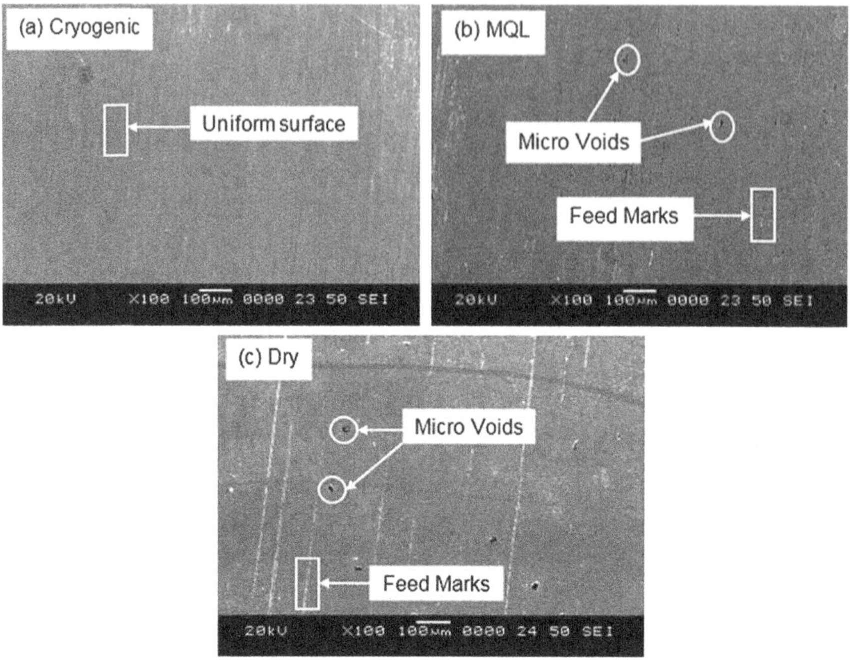


**Fig. 1** Slide burnished surface morphology

process, parameters considered in this study and the amount of lubrication splashed at the burnishing zone may not be sufficient to cause the uniform flow of the metal over the surface of the workpiece. Hence, the microvoids have not been filled in the case of MQL environments, whereas in the dry environment, the deteriorated surface was observed at a similar diamond burnishing condition. In the dry environment, the absence of lubrication causes an increase in the temperature generated in the burnishing zone. The plastic deformation on the surface layer also increases because of the sudden increase in the temperature generated. In the cooling environments except for dry conditions, the influence of excessive plastic deformation has been reduced as a reason of lubrication. These findings agree with the literature findings [25, 26].

## 3.2 Subsurface Hardness

Fatigue life and wear resistance of the product purely depends on the surface and subsurface hardness of the material. The microhardness of the specimen was measured using OMNI TECHMVHS-AUTO-type Vickers microhardness tester. Initially, the specimen was cold-mounted using acrylic powder and self-curing liquid. Further, the specimen was subjected to polishing and ultrasonic cleaning. For the

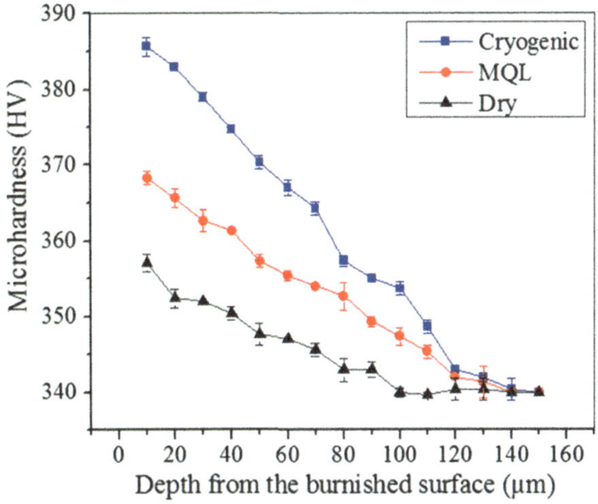


**Fig. 2** Subsurface microhardness of the burnished sample under cryogenic, MQL, and dry environment

measurement of subsurface microhardness, dwell time of 15 s and a load of 10 kgf were considered. Figure 2 represents the microhardness of the specimen which has been measured at a burnishing speed = 47 m/min, burnishing force = 90 N, and burnishing feed = 0.055 mm/rev under different lubricants. The point just below the burnished surface has been represented by 0 μm. The measurement has been carried out till the depth of 150 μm from the top surface layer of the specimen. At this point, the subsurface microhardness of the specimen reached the bulk hardness of the material. The bulk hardness of the material before diamond burnishing was found to be 340 HV. It was observed that the microhardness declines continuously from the top surface layer of the specimen. Finally, the microhardness of the specimen reaches the value of the bulk hardness of the material. It was also noticed that a similar trend has been observed in all three environments. The reason behind this variation may be explained by the fact that after performing diamond burnishing under varying lubrication conditions, a lower strain rate and less shearing will be produced [27, 28]. The cryogenic environment has produced excellent microhardness improvement in contrast with the other two environments, and also, the highest microhardness was observed for the cryogenic environment. The percentage improvement recorded in the case of cryogenic burnishing was 5% and 8% contrasted with MQL and dry environment, respectively. The percentage of improvement was noticed due to the grain refinement which is promoted by the application of $LN_2$ [26].

### 3.3 *Residual Stress*

It is believed that the residual stresses induced after burnishing have an impact on the tribological properties and fatigue life of the component. Figure 3 depicts the distribution of residual stress under all three environments. The residual stresses

**Fig. 3** Residual stress distribution under varying diamond burnishing environments

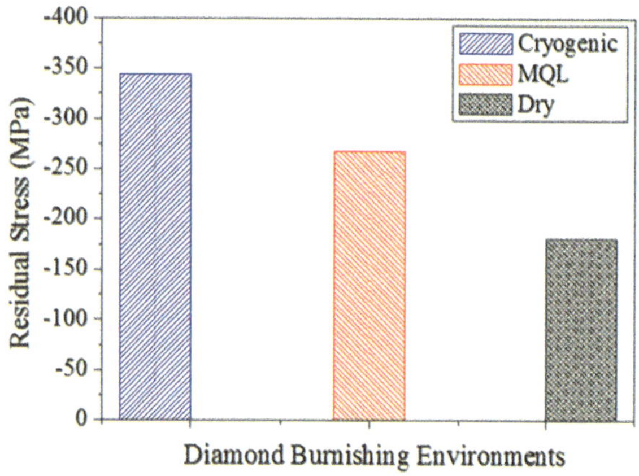


were measured at a burnishing speed = 47 m/min, burnishing feed = 0.055 mm/rev, and burnishing force = 90 N under three environments. It was observed that after the diamond burnishing process compressive residual stresses had been induced on the surface under three environments. Compressive residual stresses of 345 MPa, 268 MPa, 181 MPa were induced under cryogenic, MQL, and dry environments, respectively. The induced compressive residual stresses on the surface show that cryogenic diamond burnishing is a better mode of lubrication for inducing favorable compressive residual stresses on the surface [7, 25]. The fatigue strength of the material is expected to be improved since the induced compressive residual stresses retard the formation of cracks which is usually developed after the burnishing process. The combination of severe plastic deformation and the splashing of the $LN_2$ at the burnishing zone is the primary reason for the inducement of maximum compressive residual stresses on the surface of the diamond burnished specimen under a cryogenic environment [4].

## *3.4 Influence of Lubrication on Diamond Burnishing Process*

The use of lubricants to avoid the formation of defects on the workpiece surface is a major concern in the industry sectors. Many researchers have worked on improving the surface characteristics of the material by the application of different kinds of lubrication at the tool–workpiece interface. In recent times, a variety of lubricants have been introduced to reduce the effect of machining on the workpiece. Coolants such as cryogenic, MQL, and flood cooling are widely used in different types of machining. Handling these lubricants causes side effects on human health and the environment. Disposal of these lubricants is also a major point of concern. Liquid nitrogen causes a cold burn on the skin [23]. The major advantage of $LN_2$ is that it gets evaporated quickly during machining without causing environmental problems. One of the major issues with MQL is that it causes an inhalation problem [27]. In both the sustainable environments, the surface characteristics and productivity were found to be improved as a reason of the constant spraying of the mist and jet at the contact zone [24]. In MQL and cryogenic environments, surface integrity was enhanced. Whereas in the dry condition due to the absence of lubrication, the result observed was not good in contrast with the other two environments. While working under the MQL environment, a low quantity of lubricant in the form of mist will be sprayed at the burnishing zone. This process reduces the developed temperature in the working zone. In a dry condition, lubrication will not be used which results in temperature rise at the tool–workpiece interface. Another major issue is that friction will be more at the contact point and also the thermal softening effect will deteriorate the surface of the material [25].

## 4 Conclusions

In the article, the importance of a sustainable cooling environment in the diamond burnishing of the material under consideration was analyzed. The key findings of the study are as follows:

- An improvement in the microhardness of 5% and 8% has been achieved in cryogenic cooling in contrast with MQL and dry environment, respectively.
- Uniform surface was noticed in the cryogenic slide burnishing.
- Compressive residual stresses have been induced on the diamond burnished surface due to the plastic deformation and grain refinement of the material.
- It has been proved that the cryogenic environment yields enhanced surface integrity characteristics of 17-4 PH stainless steel after diamond burnishing. It is recommended to use cryogenic cooling to improve the surface integrity characteristics of the material.

## References

1. Lyons AC, Ne M (2000) An investigation of the surface topography of ball burnished mild steel and aluminium. Int J Adv Manuf Technol 469–473
2. Mohanty A, Gangopadhyay S, Thakur A (2016) On applicability of multilayer coated tool in dry machining of aerospace grade stainless steel. Mater Manuf Process. 31:869–879. https://doi.org/10.1080/10426914.2015.1070413
3. Sachin B, Narendranath S, Chakradhar D (2020) Application of desirability approach to optimize the control factors in cryogenic diamond burnishing. Arab J Sci Eng:1–13.
4. Sachin B, Narendranath S, Chakradhar D (2019) Enhancement of surface integrity by cryogenic diamond burnishing toward the improved functional performance of the components. J Braz Soc Mech Sci Eng 41(10):396

5. Klim Z, Ennajimi E, Balazinski M, Fortin C (1996) Cutting tool reliability analysis for variable feed milling of 17-4 PH stainless steel. Wear 195:206–213. https://doi.org/10.1016/0043-1648(95)06863-5
6. Sachin B, Narendranath S, Chakradhar D (2019) Analysis of surface hardness and surface roughness in diamond burnishing of 17-4 PH stainless steel. In: IOP conference series: materials science and engineering, vol 577. IOP Publishing; 012075. No. 1
7. Sachin B, Narendranath S, Chakradhar D (2019) Effect of working parameters on the surface integrity in cryogenic diamond burnishing of 17–4 PH stainless steel with a novel diamond burnishing tool. J Manuf Process 38:564–571
8. Toboła D, Jerzy M, Łukasz M (2020) TEM analysis of surface layer of Ti–6Al–4V ELI alloy after slide burnishing and low-temperature gas nitriding. Appl Surf Sci 145942
9. Świrad S (2011) The surface texture analysis after sliding burnishing with cylindrical elements. Wear 271(3–4):576–581
10. Korzynska K, Zeglicki W, Korzynski M, Drabczyk M (2020) Flow burnishing with glass microspheres and its impact on the surface condition of 2017A alloy elements. J Manuf Process 58:1297–1303
11. Banh QN, Shiou FJ (2016) Determination of optimal small ball-burnishing parameters for both surface roughness and superficial hardness improvement of STAVAX. Arab J Sci Eng 41:639–652. https://doi.org/10.1007/s13369-015-1710-1
12. Yildiz Y, Nalbant M (2008) A review of cryogenic cooling in machining processes. Int J Mach Tools Manuf 48:947–964. https://doi.org/10.1016/j.ijmachtools.2008.01.008
13. Sachin B, Narendranath S, Chakradhar D (2019) Selection of optimal process parameters in sustainable diamond burnishing of 17-4 PH stainless steel. J Braz Soc Mech Sci Eng 41(5):219
14. Rao CM, Rao SS, Herbert MA (2019) Studies on the effect of process parameters in turning of Ti-6Al-4V alloy using topsis. In: IOP conference series: materials science and engineering, vol 577. IOP Publishing, 012069. No. 1
15. Rao CM, Rao SS, Herbert MA (2020) An experimental and numerical approach to study the performance of modified perforated cutting tools on machining of Ti–6Al–4V alloy. Arab J Sci Eng 45(2):1191–206
16. Rao CM, Rao SS, Herbert MA (2018) Development of novel cutting tool with a micro-hole pattern on PCD insert in machining of titanium alloy. J Manuf Process 36:93–103. https://doi.org/10.1016/j.jmapro.2018.09.028
17. Bordin A, Sartori S, Bruschi S, Ghiotti A (2016) Experimental investigation on the feasibility of dry and cryogenic machining as sustainable strategies when turning Ti6Al4V produced by additive manufacturing. J Clean Prod 142:4142–4151. https://doi.org/10.1016/j.jclepro.2016.09.209
18. Rao CM, Rao SS, Herbert MA (2018) Influence of modified cutting inserts in machining of Ti-6Al-4V alloy using PCD insert. Mater Today Proc 5:18426–32. https://doi.org/10.1016/J.MATPR.2018.06.183
19. Vishwas CJ, Gajanan MN, Sachin B, Abhinaba R, Puneet NP, Anjan BN, Vinayak NK (2019) Study on surface roughness in minimum quantity lubrication turning of Al-6082/SiC metal matrix composites. Appl Mechan Mater 895:127–33. Trans Tech Publications Ltd
20. Vishwas CJ, Girish LV, Naik GM, Sachin B, Roy A, Prashanth BY, Badiger R (2018) Effect of machining parameters on surface integrity during dry turning of AISI 410 martensitic stainless steel. IOP Conf Ser Mater Sci Eng 376
21. Charitha R, Shrikantha R, Mervin H (2018) Performance improvement studies for cutting tools with perforated surface in turning of titanium alloy. MATEC web of conferences. EDP Sciences, 144
22. Tang J, Luo HY, Zhang YB (2016) Enhancing the surface integrity and corrosion resistance of Ti-6Al-4V titanium alloy through cryogenic burnishing. Int J Adv Manuf Technol 1–9. https://doi.org/10.1007/s00170-016-9000-y
23. Huang B, Kaynak Y, Sun Y, Jawahir IS (2015) Surface layer modification by cryogenic burnishing of Al 7050–T7451 alloy and validation with FEM-based burnishing model. Procedia CIRP 31:1–6. https://doi.org/10.1016/j.procir.2015.03.097

24. Yang S, Umbrello D, Dillon OW, Puleo DA, Jawahir IS (2015) Cryogenic cooling effect on surface and subsurface microstructural modifications in burnishing of Co-Cr-Mo biomaterial. J Mater Process Technol 217:211–221. https://doi.org/10.1016/j.jmatprotec.2014.11.004
25. Sachin B, Narendranath S, Chakradhar D (2019) Sustainable diamond burnishing of 17–4 PH stainless steel for enhanced surface integrity and product performance by using a novel modified tool. Mater Res Express 6(4):046501
26. Sachin B, Narendranath S, Chakradhar D (2018) Experimental evaluation of diamond burnishing for sustainable manufacturing. Mater Res Expr 5(10):106514
27. Rao CM, Sachin B, Rao SS, Herbert MA (2020) Minimum quantity lubrication through the micro-hole textured PCD and PCBN inserts in the machining of the Ti–6Al–4V alloy. Tribol Int 106619
28. Sachin B, Narendranath S, Chakradhar D (2018) Effect of cryogenic diamond burnishing on residual stress and microhardness of 17-4 PH stainless steel. Mater Today Proc 5(9):18393–18399

# Investigation of Effect of EDM Process Variables on Material Removal Rate and Tool Wear Rate in Machining of EN19 Steel Using Response Surface Methodology

**Santosh Nandurkar, Sachin Kulkarni, Tushar Hawal, Niranjan Pattar, and Nagaraj Kelageri**

**Abstract** The work presented here gives the details of effect of process variables, namely current, pulse-on time and pulse-off time on material removal rate and tool wear rate in electrical discharge machining (EDM) of hardened EN19 steel. The copper tool electrode was used for experimentation. To study the effect of parameters on process characteristics, response surface methodology and Box-Behnken experimental design were employed. The models for responses were developed based on experimental results. Confirmative tests were conducted to validate the results, and less than 5% error was observed between predicted and experimental values of the responses. At the end, it has been found that the current has major effect on responses of EDM (material removal rate and tool wear rate), whereas there are some interaction effects possible between current and pulse-off time on material removal rate and tool wear rate.

**Keywords** Hardened · Machining · Response surface methodology · Die-sink EDM

## 1 Introduction

A non-conventional machining process in which material is removed by thermal erosion (melting and evaporation technique) is called electrical discharge machining (EDM). Extremely hard materials with low machinability can be machined using a non-conventional thermoelectric source of energy. EDM is greatly helpful in achieving economical machining of hard materials. In EDM, a series of electric sparks [1] through controlled erosion removes the material from the workpiece. EDM is best manufacturing method for accurately producing intricate shapes and complex

S. Nandurkar (✉) · N. Pattar · N. Kelageri
KLE Dr M S Sheshgiri College of Engineering and Technology, Belgaum 590008, India

S. Kulkarni · T. Hawal
KLS Gogte Institute of Technology, Belgaum 590008, India



P. Srinivasa Pai and V. Krishnaraj (eds.), *Sustainable Machining Strategies for Better Performance*, Lecture Notes in Mechanical Engineering,
https://doi.org/10.1007/978-981-16-2278-6_7

geometries in extremely hard materials. High number of sparks per second is generated to produce small craters in the material through melting and evaporation, which causes erosion of workpiece as well as tool electrode [2]. The temperature generated during each spark ranges from 8000 to 12,000 °C [3] and as high as 20,000 °C [4]. The material removal is dependent of on various factors such as current, pulse duration, voltage, discharge frequency, type of workpiece and tool electrode, dielectric fluid used for flushing, flushing conditions. Tool electrode should be highly electrically and thermally conductive with superior properties like wear resistance and high melting point. Hard materials having low machinability but good electrically conductivity can be machined using EDM. In this study, the focus is given on machining of EN 19 alloy steel with hardness of 55-60HRC, whose applications in making machine tools, gears, bolts and studs are growing rapidly. The parametric effect over the removal of material and tool erosion is studied. EDM has become an important precision machine tool to produce an internal shape on workpieces which are conventionally difficult to produce.

Many studies have been carried out on electrical discharge machining of harder material to produce simple and complex shapes. Shabgard and Co-authors [5] have studied the parametric effect on the characteristics of EDM process. They had used full factorial design in the machining of AISI H13 tool steel. Guu and Hocheng [6] worked on effect of workpiece rotations on EDM of AISI D2 tool steel with copper as tool electrode. Khan and Hameedullah [7] investigated the effect of tool polarity on the machining of silver steel. The effect of copper and aluminum with kerosene as dielectric medium was studied by Pradhan and Jayswal [8]. Marafona [9] has studied the black layer formation of on the workpiece and tool electrode when machining was carried out. Pradhan and Biswas [10] used the response surface methodology to establish relations and interaction effects between the process parameters and material removal rate (MRR). Kansal and Co-authors [11] carried out EDM on EN 31 tool steel using copper electrode. Asif Iqbal and Khan [12] conducted study on influence of controlled variables in EDM of AISI 304 stainless steel. From the literature, it has been seen that very few authors have worked on the investigation of effect of copper tool electrode on hardened EN 19 alloy steel. So this study focused on effect of main parameters of EDM on MRR and tool wear rate (TWR) while machining of EN 19 steel using copper as tool material.

## 2 Materials, Methodology and Experimentation

In this section, workpiece and tool electrode materials along with response surface methodology have been discussed. Die-sink-type electrical discharge machine was used to perform the experiments on oil-hardened EN 19 steel using copper as tool electrode material. The size of the workpiece used was having a diameter of 60 mm and thickness (height) of 8 mm, and it was composed of 97.15% iron, 0.4% carbon, 0.7% manganese, 1.2% chromium, 0.25% silicon and 0.3% molybdenum. The copper tool electrode used was 20 mm in diameter and 100 mm long. It was composed of

99.95% copper, 0.03% $O_2$ and metallic impurities. Kerosene was used as a dielectric fluid. Figure 1 shows the die-sink-type electrical discharge machine of ELECTRA pride-Z make, and Fig. 2 shows the workpiece and tool electrode materials with dimensions.

EN 19 steel had a hardness of 55-60HRC after oil quenching, and density was 7770 kg/m$^3$. Before conducting the EDM, top and bottom surfaces of EN 19 were smoothly finished using surface grinding machine. The bottom face of tool electrode was also cleaned and polished for smooth finish. A weighing machine of KEROY make with 0.05 g accuracy, and weighing range from 0.05 to 300 g was used to weigh the workpiece and tool electrode. While experimentation, workpiece was connected to anode, and tool electrode was connected to cathode. The machining time ($T$) was fixed to 20 min for each experiment. After each experiment, the difference in weight of the workpiece and tool electrode was calculated by weighing before and after machining to find out MRR and TWR. The required data was collected after each experiment till all experiments were conducted.

The process variables considered here were current, pulse-on time and pulse-off time. The variables and their levels are given in Table 1.

The responses or process characteristics considered here were MRR and TWR. Equations 1 and 2 give the MRR and TWR.

**Fig. 1** Die-sink electrical discharge machine

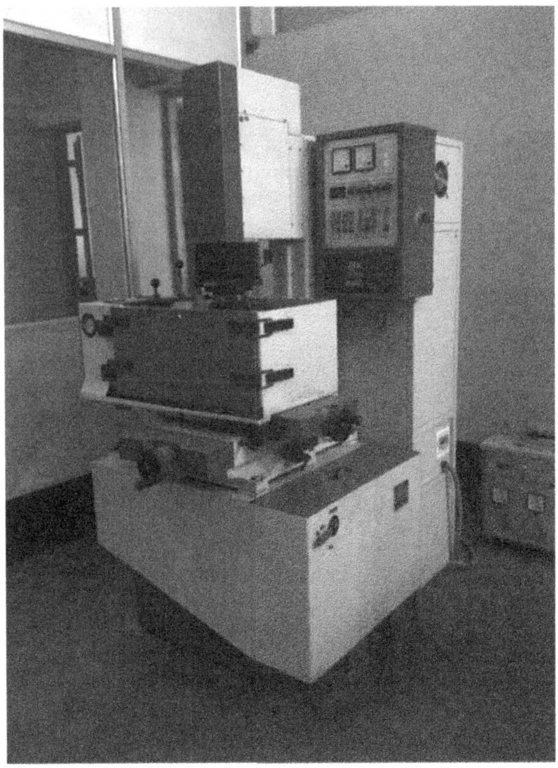

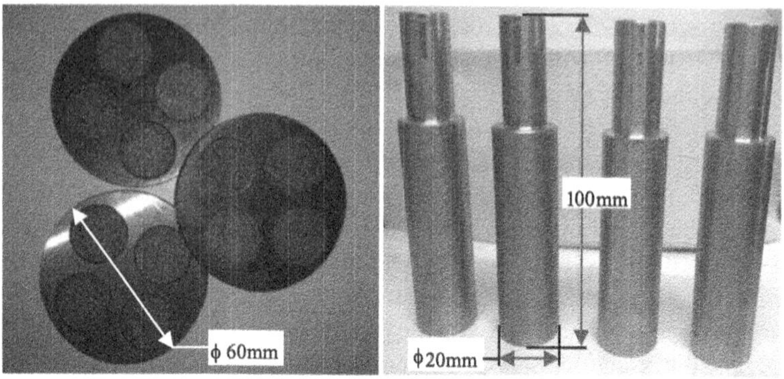


(a) (b)

**Fig. 2** **a** Machined workpieces of EN 19 steel, **b** copper tool electrodes

**Table 1** Process variables and their levels

| Sl. No. | Process variables | Uncoded values | Coded values | Levels | | | Ranges |
|---|---|---|---|---|---|---|---|
| | | | | Low (−1) | Mid (0) | High (+1) | |
| 1 | Current | $I$ | $X_1$ | 8 | 14 | 20 | 1–20 A |
| 2 | Pulse-on time | $T_{on}$ | $X_2$ | 60 | 79 | 98 | 0–99 μs |
| 3 | Pulse-off time | $T_{off}$ | $X_3$ | 5 | 7 | 9 | 0–9 μs |

$$\text{MRR} = \frac{\text{Amount of material removed from workpiece}}{\text{Machining Time } (T)} \tag{1}$$

$$\text{TWR} = \frac{\text{Amount of material removed from tool electrode}}{\text{Machining Time } (T)} \tag{2}$$

The experiments were designed using Box-Behnken Design (BBD) technique under response surface methodology (RSM). The design matrix for carrying out experiments was given by BBD by taking points at the center of each edge of the cube and three at the center of the cube. Table 2 gives the set of experiments and corresponding values of MRR and TWR.

In Table 2, run order gives the sequence of experiments, the uncoded values represent the actual values of process variables, and coded values give the low (−1), mid (0) and high (+1) levels of each process variables. The levels were calculated using following Eq. 3.

$$X_i = \frac{2X - (X_{max} + X_{min})}{X_{max} - X_{min}} \tag{3}$$

**Table 2** Set of experiments and corresponding values of MRR and TWR

| Run order | Uncoded values | | | Coded values | | | $Y_1 = \text{MRR}$ (gm/min) | $Y_2 = \text{TWR}$ (gm/min) |
|---|---|---|---|---|---|---|---|---|
| | $I$ (A) | $T_{on}$(μs) | $T_{off}$(μs) | $X_1$ | $X_2$ | $X_3$ | | |
| 1 | 14 | 79 | 7 | 0 | 0 | 0 | 0.295 | 0.0040 |
| 2 | 14 | 98 | 5 | 0 | +1 | −1 | 0.265 | 0.0023 |
| 3 | 20 | 60 | 7 | +1 | −1 | 0 | 0.375 | 0.0050 |
| 4 | 20 | 79 | 5 | +1 | 0 | −1 | 0.420 | 0.0039 |
| 5 | 8 | 98 | 7 | −1 | +1 | 0 | 0.130 | 0.0010 |
| 6 | 14 | 60 | 9 | 0 | −1 | +1 | 0.266 | 0.0060 |
| 7 | 20 | 79 | 9 | +1 | 0 | +1 | 0.449 | 0.0058 |
| 8 | 8 | 79 | 9 | −1 | 0 | +1 | 0.156 | 0.0025 |
| 9 | 14 | 98 | 9 | 0 | +1 | −1 | 0.273 | 0.0028 |
| 10 | 8 | 60 | 7 | −1 | −1 | 0 | 0.156 | 0.0023 |
| 11 | 8 | 79 | 5 | −1 | 0 | −1 | 0.138 | 0.0025 |
| 12 | 14 | 79 | 7 | 0 | 0 | 0 | 0.294 | 0.0035 |
| 13 | 20 | 98 | 7 | +1 | +1 | 0 | 0.365 | 0.0030 |
| 14 | 14 | 60 | 5 | 0 | −1 | −1 | 0.255 | 0.0039 |
| 15 | 14 | 79 | 7 | 0 | 0 | 0 | 0.298 | 0.0034 |

where

$X_i$ is coded value for variable $X$ and $i = 1, 2, 3$ for $I$, $T_{on}$ and $T_{off}$, respectively.

$X$ is any value of the variable from $X_{max}$ to $X_{min}$.

$X_{max}$ and $X_{min}$ are lower and upper levels of the variable $X$.

The regression coefficients of the models were found out using following Eq. 4.

$$b = \left(X^T X\right)^{-1} X^T Y \tag{4}$$

where

$b$ is matrix of variable estimates

$X$ is calculation matrix

$Y$ is matrix of measured response.

Response surface methodology was employed to find the mathematical relationship between the response ($Y_n$) and process variables. Equation 5 gives the generalized form of second-order polynomial response surface model.

$$Y_n = b_o + \sum_{i=1}^{3} b_i X_i + \sum_{i=1}^{3} b_{ii} X_i^2 + \sum_{i=1}^{2} \sum_{j=i+1}^{3} b_{ij} X_i X_j \tag{5}$$

where

$Y_n$ is response under study ($n = 1$ for MRR and $n = 2$ for TWR).

$X_i$ is coded value for variable $X$ and $i = 1, 2, 3$ for $I$, $T_{on}$ and $T_{off}$, respectively.

$b_o, b_i, b_{ii}, b_{ij}$ are second-order regression coefficients.

The significant terms were decided based on 95% confidence level, and corresponding coefficients were obtained from Table 3 for $Y_1$ and $Y_2$. Minitab [13] software version 16.0 was used to find regression coefficients of response models for MRR and TWR.

Equations 6 and 7 represent MRR and TWR, respectively.

$$Y_1 = 0.295583 + 0.128594X_1 - 0.032510X_2^2 \tag{6}$$

$$Y_2 = 0.003633 + 0.001169X_1 - 0.001019X_2 - 0.000562X_3 + 0.000471X_3^2 + 0.000463X_1X_3 \tag{7}$$

The Minitab output for analysis of variance (ANOVA) is given in Tables 4 and 5 along with values of coefficient of multiple determination (R-sq).

To validate the above models, confirmative tests were conducted, and it has been found that the error in experimental and predicted value is less than 5%. Tables 6 and 7 represent the details of confirmative tests, and the error was calculated using following formula.

$$\text{Error (\%)} = \frac{\text{Predicted Value} - \text{Experimental Value}}{\text{Predicted Value}} \times 100 \tag{8}$$

**Table 3** Estimation of regression coefficients for $Y_1$ = MRR and $Y_2$ = TWR

| Terms | | For $Y_1$ = MRR | | For $Y_2$ = TWR | |
|---|---|---|---|---|---|
| | | Coefficients | *P*-value | Coefficients | *P*-value |
| Constant | – | 0.295583 | 0.000* | 0.003633 | 0.000* |
| Current | $X_1$ | 0.128594 | 0.000* | 0.001169 | 0.000* |
| Pulse-on time | $X_2$ | −0.002469 | 0.757 | −0.001019 | 0.000* |
| Pulse-off time | $X_3$ | 0.008375 | 0.319 | 0.000562 | 0.006* |
| Current × current | $X_1 \times X_1$ | −0.006510 | 0.584 | −0.000442 | 0.056 |
| Pulse-on time × pulse-on time | $X_2 \times X_2$ | −0.032510 | 0.033* | −0.000367 | 0.095 |
| Pulse-off time × pulse-off time | $X_3 \times X_3$ | 0.001677 | 0.886 | 0.000471 | 0.046* |
| Current × pulse-on time | $X_1 \times X_2$ | 0.004062 | 0.720 | -0.000175 | 0.354 |
| Current × pulse-off time | $X_1 \times X_3$ | 0.002625 | 0.816 | 0.000463 | 0.043* |
| Pulse-on time × pulse-off time | $X_2 \times X_3$ | −0.000750 | 0.947 | −0.000388 | 0.073 |

*Significant terms and these terms were found out from the *P*-value which is less than or equal to 0.05 for 95% confidence level

**Table 4** Analysis of variance for response model of $Y_1$ = MRR

| Source | DF | Seq SS | Adj SS | Adj MS | F | P |
|---|---|---|---|---|---|---|
| Regression | 9 | 0.137033 | 0.137033 | 0.015226 | 33.23 | 0.001 |
| Linear | 3 | 0.132901 | 0.132901 | 0.044300 | 96.70 | 0.000 |
| Current | 1 | 0.132901 | 0.132291 | 0.132291 | 288.76 | 0.000 |
| Pulse-on time | 1 | 0.000049 | 0.000049 | 0.000049 | 0.11 | 0.757 |
| Pulse-off time | 1 | 0.000561 | 0.000561 | 0.000561 | 1.22 | 0.319 |
| Square | 3 | 0.004037 | 0.004037 | 0.001346 | 2.94 | 0.138 |
| Current × current | 1 | 0.000069 | 0.000157 | 0.000157 | 0.34 | 0.584 |
| Pulse-on time × pulse-on time | 1 | 0.003957 | 0.003903 | 0.003903 | 8.52 | 0.033 |
| Pulse-off time × pulse-off time | 1 | 0.000010 | 0.000010 | 0.000010 | 0.02 | 0.886 |
| Interaction | 3 | 0.000096 | 0.000096 | 0.000032 | 0.07 | 0.974 |
| Current × pulse-on time | 1 | 0.000066 | 0.000066 | 0.000066 | 0.14 | 0.720 |
| Current × pulse-off time | 1 | 0.000028 | 0.000028 | 0.000028 | 0.06 | 0.816 |
| Pulse-on time × pulse-off time | 1 | 0.000002 | 0.000002 | 0.000002 | 0.00 | 0.947 |
| Residual error | 5 | 0.002291 | 0.00291 | 0.000458 | | |
| Lack of fit | 3 | 0.002284 | 0.002284 | 0.000761 | 212.42 | 0.005 |
| Pure error | 2 | 0.000007 | 0.000007 | 0.000004 | | |
| Total | 14 | 0.139324 | | | | |
| *R*-Sq = 98.36% *R*-Sq{adj} = 95.40% | | | | | | |

## 3 Results and Discussion

In this section, effect of different variables on MRR and TWR has been studied. Figure 3 illustrates the effect of current on MRR and TWR. It has been observed that as current increases, the MRR increases [14, 15]. Increase in current also increases TWR. This happened due to the melting of material from surfaces of workpiece and tool electrode at high current values. Figure 4 shows relationship between pulse-on time and responses (MRR and TWR) where MRR increases till mid-level and then drops suddenly. This occurred due to the choking of wear particles during high pulse-on time. In contrast to this, decrease in TWR was observed with increase in pulse-on time that happened due to less number of wear particles hitting the tool surface. There was no significant effect of pulse-off time on MRR [14, 15] but it showed effect on TWR, and Fig. 5 depicts the same. As there was increase in pulse-off time, there was gradual increase in TWR in the initial stage and sudden increase took place with further increase in pulse-off time. Also from Tables 4 and 5, it was observed that for MRR, the value of R-sq is 98.36%, and R-sq{adj} is 95.40% which is well within the acceptable limits. Similarly, for TWR, the value of R-sq is 97.75%, and R-sq{adj} is 93.70% that shows the good agreement with obtained results.

**Table 5** Analysis of variance for response model of $Y_2$ = TWR

| Source | DF | Seq SS | Adj SS | Adj MS | F | P |
|---|---|---|---|---|---|---|
| Regression | 9 | 0.000026 | 0.000026 | 0.000003 | 24.14 | 0.001 |
| Linear | 3 | 0.000022 | 0.000022 | 0.000007 | 61.76 | 0.000 |
| Current | 1 | 0.000011 | 0.000011 | 0.000011 | 93.04 | 0.000 |
| Pulse-on time | 1 | 0.000008 | 0.000008 | 0.000008 | 70.69 | 0.000 |
| Pulse-off time | 1 | 0.000003 | 0.000003 | 0.000003 | 21.55 | 0.006 |
| Square | 3 | 0.000002 | 0.000002 | 0.000001 | 6.17 | 0.039 |
| Current × current | 1 | 0.000001 | 0.000001 | 0.000001 | 6.13 | 0.056 |
| Pulse-on time × pulse-on time | 1 | 0.000001 | 0.000000 | 0.000000 | 4.23 | 0.095 |
| Pulse-off time × pulse-off time | 1 | 0.000001 | 0.000001 | 0.000001 | 6.97 | 0.046 |
| Interaction | 3 | 0.000002 | 0.000002 | 0.000001 | 4.48 | 0.070 |
| Current × pulse-on time | 1 | 0.000000 | 0.000000 | 0.000000 | 1.04 | 0.354 |
| Current × pulse-off time | 1 | 0.000001 | 0.000001 | 0.000001 | 7.28 | 0.043 |
| Pulse-on time × pulse-off time | 1 | 0.000001 | 0.000001 | 0.000001 | 5.11 | 0.073 |
| Residual error | 5 | 0.000001 | 0.000001 | 0.000000 | | |
| Lack of fit | 3 | 0.000000 | 0.000000 | 0.000000 | 1.23 | 0.478 |
| Pure error | 2 | 0.000000 | 0.000000 | 0.000000 | | |
| Total | 14 | 0.000026 | | | | |
| *R*-Sq = 97.75% *R*-Sq{adj} = 93.70% | | | | | | |

# 4 Conclusion

From this study, it has been concluded that using RSM and Box-Behnken design, accurate response models can be developed to predict the MRR and TWR in EDM process. The results show that current is the most significant variable compared to pulse-on time and pulse-off time for MRR. In contrast to this for TWR, the current, pulse-on time and pulse-off time prove to be significant variables. It can also be concluded that higher values of MRR can be achieved at a higher value of current but simultaneously, the wear of the tool electrode also increases.

**Table 6** Confirmative tests for $Y_1$ = MRR

| Run order | Uncoded values | | | Coded values | | | MRR (gm/min) | | |
|---|---|---|---|---|---|---|---|---|---|
| | $I$ (A) | $T_{on}$(μs) | $T_{off}$(μs) | $X_1$ | $X_2$ | $X_3$ | Predicted values | Experimental values | Error (%) |
| 1 | 10 | 65 | 7 | −2/3 | −14/19 | 0 | 0.1954 | 0.2018 | −3.27 |
| 2 | 12 | 75 | 9 | −1/3 | −4/19 | +1 | 0.2491 | 0.2522 | −1.24 |
| 3 | 16 | 85 | 6 | 1/3 | 6/19 | −1/2 | 0.3348 | 0.3195 | 4.57 |
| 4 | 18 | 95 | 8 | 2/3 | 16/19 | 1/2 | 0.3668 | 0.3843 | −4.77 |

**Table 7** Confirmative tests for $Y_2$ = TWR

| Run order | Uncoded values | | | Coded values | | | TWR (gm/min) | | |
|---|---|---|---|---|---|---|---|---|---|
| | $I$ (A) | $T_{on}$(µs) | $T_{off}$(µs) | $X_1$ | $X_2$ | $X_3$ | Predicted values | Experimental values | Error (%) |
| 1 | 10 | 65 | 7 | −2/3 | −14/19 | 0 | 0.0036 | 0.0035 | 2.77 |
| 2 | 12 | 75 | 9 | −1/3 | −4/19 | +1 | 0.0043 | 0.0042 | 2.32 |
| 3 | 16 | 85 | 6 | 1/3 | 6/19 | −1/2 | 0.0034 | 0.0035 | -2.94 |
| 4 | 18 | 95 | 8 | 2/3 | 16/19 | 1/2 | 0.0041 | 0.0040 | 2.44 |

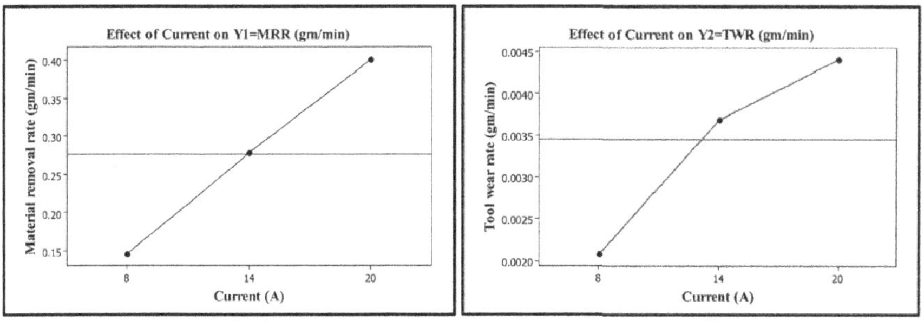


(a) (b)

**Fig. 3** Effect of current on **a** MRR and **b** TWR

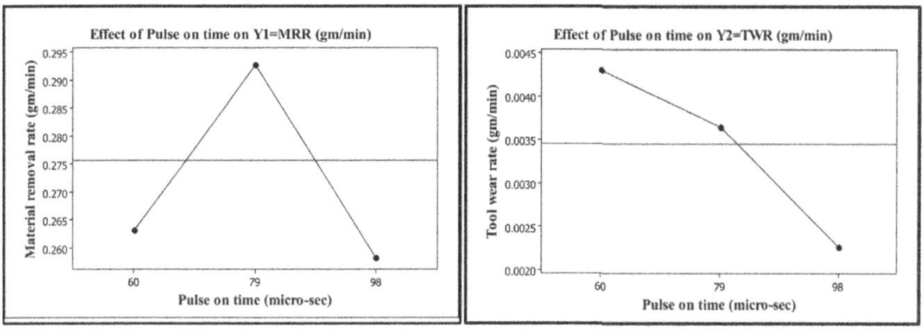


(a) (b)

**Fig. 4** Effect of pulse-on time on **a** MRR and **b** TWR

**Fig. 5** Effect of pulse-off time on TWR

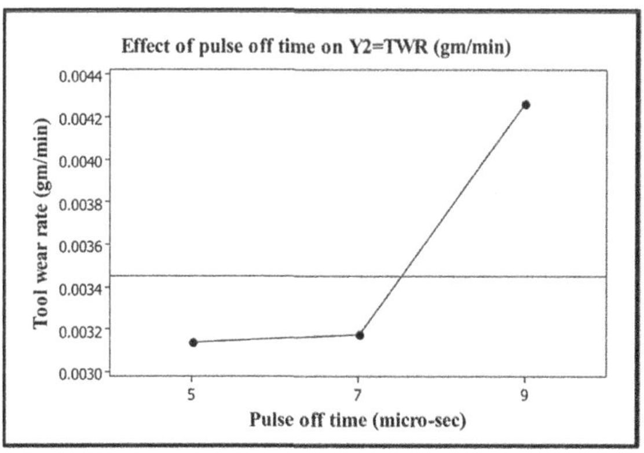

## References


1. Ghosh A, Mallick AK (2006) Manufacturing science. Affiliated East-West press Pvt. Ltd., New Delhi
2. Kalpakjian S, Schmid SR, Manufacturing processes for engineering materials, 5th edn. Dorling Kindersley (India) Pvt. Ltd, Delhi
3. Konig W, Dauw DF, Levy G, Panten U (1988) EDM-Future steps towards the machining of ceramics. Ann CIRP 37(2):623–631
4. Boothroyd G, Winston AK (1989) Nonconventional machining processes: fundamentals of machining and machine tools. Marcel Dekker, New York
5. Shabgard M, Seyedzavvar M, Oliaei SNB (2011) Influence of input parameters on the characteristics of the EDM Process. J Mech Eng 57(9):689–696
6. Guu YH, Hocheng H (2001) Effects of work piece rotation on machinability during electrical discharge machining. J Mater Manuf Process 16:91–101
7. Khan DA, Hameedullah M (2011) Effect of tool polarity on the machining characteristics in electric discharge machining of silver steel and statistical modelling of the process. Int J Eng Sci Technol 3:5001–5010
8. Pradhan D, Jayswal SC (2011) Behaviour of copper and aluminium electrodes on EDM of EN-8 alloy steel. Int J Eng Sci Technol 3:5492–5499
9. Marafona J (2007) Black layer characterisation and electrode wear ratio in electrical discharge machining (EDM). J Mater Process Technol 184:27–31
10. Pradhan MK, Biswas CK (2008) Modelling of machining parameters for MRR in EDM using response surface methodology. In: Proceedings of national conference on mechanism science and technology: from theory to application, 535–542
11. Kansal HK, Singh S, Kumar P (2005) Parametric optimization of powder mixed electrical discharge machining by response surface methodology. J Mater Process Technol 169:427–436
12. Asif Iqbal AKM, Khan AA (2010) Influence of process parameters on electrical discharge machined job surface integrity. Am J Eng Appl Sci 3:396–402
13. Ward SM, Bin Z, Thouless MD (2005) Predicting the failure of ultrasonic spot welds by pull-out from sheet metal. Int J Solids Struct 43:7482–7500
14. Faisal N, Kaushik K (2018) Optimization of machine process parameters in EDM for EN 31 using evolutionary optimization techniques. Technologies 6:54
15. Regmi M, Gupta A (2017) Performance of Copper electrode for machining EN 19 and Nickel Plated EN 19 alloy steel by EDM. Int J Indus Manuf Syst Eng 2(1):1–6

# Sustainability Analysis of Cutting Fluids in Minimum Quantity Lubrication of Machining Operations

**P. Jamaleswara Kumar and B. V. S. Arun Kumar**

**Abstract** This paper is a machining investigation on the sustainability of two potential cutting fluids in a machining process using a custom built, low-cost dispensing system, based on MQL, for delivering cutting fluids. A preliminary experimental investigation was conducted on unblended oils, i.e. metal working fluids (MWFs), with a view to further develop a better effective cutting fluid along with a delivery system. Influence of basic machining parameters on the performance of the operation has been investigated to detect any significant difference in using either of the fluids on a simple plain turning operation using an HSS cutting tool. With a response variable like surface roughness, it was possible to notice significant differences in the machining outputs. Low-cost fluids have been used, and the feasibility of a low-cost enhancement of the properties is concluded. This study considers the real situation of imperfect fluids and proportion of it can be reworked at certain known rate. The overall purpose of this analysis is the development of a highly economical green cutting fluid dispenser suitable for a typical small-scale industrial manufacturer, with a job shop-type setup, based on a low-cost investigation. Both the considered fluids have been evaluated based on past research and found to be good starting points for further study. The vegetable oil was found to perform better on machining outputs but score close to its counterpart on sustainability.

**Keywords** Sustainable manufacturing · Metal working fluids (MWFs) · Minimum quantity lubrication (MQL) · Surface roughness · Orthogonal array and Taguchi technique

## 1 Introduction

Although Dornfeld et al. [1] emphasized a balance in the approach towards bringing sustainability, sometimes economical factors far outweigh the other two factors

P. Jamaleswara Kumar (✉) · B. V. S. Arun Kumar
Department of Mechanical Engineering, Raghu Engineering College, Visakhapatnam, India

 
P. Srinivasa Pai and V. Krishnaraj (eds.), *Sustainable Machining Strategies for Better Performance*, Lecture Notes in Mechanical Engineering,
https://doi.org/10.1007/978-981-16-2278-6_8

involved in developing sustainable manufacturing methods. While green manufacturing focusses only on being environmentally benign and socially conscious, sustainability brings in economics of the manufacturing process into the picture. But at a microeconomic level, manufacturers depend on quality driven by costs to increase their competitivity.

Within the scope of manufacturing, greening of machining processes has strongly focussed over the lubrication and cooling processes which have grown from use of plain water to the most sophisticated nanofluids to maintain a sustainable balance between the growing needs of the industry—society, economics and environment. Alternatives for handling these issues, according to Srikant and Rao [2], can be dry machining, MQL, compressed gas cooling, internal cooling, tool coatings or cryogenic cooling. As a sustainability enabler, MQL in machining has been proven and developed by many researchers, to the current state of the art. The tool–work interface is a thermo-mechanical zone due to intense tribological interaction, in many conventional machining processes is a heat dissipating zone requiring some form of cooling process. Over the years, the cooling medium used in this process has transitioned from plain water to very sophisticated fluids enhanced by nanoparticles. Several authors have reported increased thermal conductivity, lubricity, enhanced convective heat transfer coefficient, lubricating properties, wettability, etc. Nano-additives like xGnP, $SiO_2$, $MoS_2$, $Al_2O_3$, ND, CNO, CNT, boric acid, CuO, Cu, graphite, etc., have been experimented with as additives with different base fluids, both vegetable oils as well as synthetic oils, in MQL machining at very high cutting speeds. Krishna et al. [3], Rao et al. [4] and Khandekar et al. [5] were some of those researchers who studied turning under the nMQL machining of AISI 1040 steel alloys with HSS or cemented carbide tools. Nano-additives reported to have reduced wetting angles along with significant decrease in cutting forces, surface roughness and coefficient of friction. For each of these nanofluids, they have determined the best MQL flow rates and additive concentration in the base fluid. Prassan et al. [6] tried electrostatic-MQL on soya bean oil by charging oil particles and improved penetrability and wettability at the cutting zone resulting in improved lubrication by increasing wetting area and promoting lubrication through the formation of a metallic oxides layer. Roy and Ghosh (2018) attempted an in-house built small quantity lubrication (SQL) technology using a twin-jet SQL system with nanofluids in high-speed turning of AISI 4140 steel with a multilayered TiN-coated carbide insert. Vishnu et al. [7] studied and analysed the performance of high-speed machining of a stronger EN-353 steel under different machining environments. They compared the performances of SAE-40 and nanoparticles of boric acid as additive. More recently Arul and Senthil Kumar [8] developed and investigated a magneto-rheological-based minimum quantity lubrication (MR-MQL) while turning Monel K500 alloy. Nanosized CuO is used as a binder in magneto-rheological fluids to lubricate the tool–chip interface. They have reported a reduction in cutting tool temperature by 9.36% and surface roughness by 26.58% when compared with the plain MQL process. Compared to dry turning and flood lubrication, MQL has sought to achieve both the above objectives by adopting a middle path. However, in case of each machining process, the flow parameters for the cutting fluid used maybe varied along with the process parameters as well as tool and work parameters. Other than these fluid properties are also likely to influence the machining performance. Vishnu et al. [7] compared performances of dry, flooded and MQL machining of EN-353 steel alloy. The performance characteristics are studied by varying cutting speed, feed rate, depth of cut, type of tool and different lubrication conditions at three levels. The cutting fluid used in machining is vegetable oil based, prepared with the composition of coconut oil, oleic acid and triethanolamine. Machining with a PVD-coated tool at 700 rpm, with a feed rate at 0.2 mm/rev and a depth of cut at 0.5 mm, was found optimal. They also reported a marginal performance of flooded lubri-cooled machining over dry and MQL machining. Jamaleswara Kumar and Venkata Vishnu (2019) in their document refers to the experimental study of the use of vegetable oil as a cutting fluid in the manufacture of EN 353 alloys to reduce the cutting temperature under different lubrication conditions, i.e. dry, floods and minimum quantitative lubrication conditions. Jamaleswara Kumar et al. (2018) in their paper provide an overview of the automotive metal models. Implementation in the automotive industry for future metallic prototyping includes redesign of a generation of two-, three- and four-wheel auto parts, mathematical simulation and software using FEM technologies, 3D printing, metal models despite 3D/4D printing, and the prototype test and research laboratories are described in detail. Finally, this review provides an overview of the future automobile industry through metallic models.

Usage of cutting fluids based on vegetable oils has been studied extensively. Srikant and Rao [2] justify the use of vegetable oils especially because of their biodegradability, low toxicity, fewer handling and disposal problems. Their contribution to increase the sustainability of the machining processes is being enhanced from several angles. Odusote and Kolawule [9] reiterated about the suitability of vegetable oil-based cutting fluids in a drilling operation. For many spindle speeds, depths of cut and feed rates, they investigated the properties and their changes in properties like lubricity, cooling ability, viscosity and wetting ability. According to them, vegetable oil-based cutting fluids can serve as suitable alternative replacement for toxic, non-biodegradable conventional petroleum-based cutting fluids. As per the investigation of Kamata and Obikawa [10], using MQL improved surface finish over a dry environment. But it was also observed that though fluid quantity resulted in a better tool life, the surface finish was not affected. Odusote and Kolawole (2013) compared the performance of palm oil and groundnut oil with a mineral oil-based cutting fluid while machining mild steel. They found groundnut oil with lowest workpiece temperature, similar to conventional oil. Overall highest chip thickness as well as improved surface roughness was found in case of palm oil followed by groundnut oil. Sen et al. [11] reviewed previously published research articles on MQL based machining. From conventional lubri-cooling agents to different vegetable oils with nano-enhancements to machine engineering alloys, in MQL systems superior machining performances have been reported, leading to higher preference over conventional systems. Research on MQL technology has covered coolant delivery systems and their design, types of coolants, cutting tool redesign and fluid impingement locations. Compared to what Astakhov [12] said that due to low penetrability, boundary lubrication on the tool–workpiece interfaces was not possible leading to negligible cooling action due to small droplet evaporation, Wang and Zhang [13] indicated the possibility that at high temperatures of tool and workpiece, polar molecules present in the vegetable oil can react with the workpiece (metal) and produce a metallic soap film, which is responsible for good lubrication property of vegetable oil. However, Astakhov's issue of low penetrability of a lubricant into the tool–work interface zone was better addressed through channels made through the tool body carrying the cutting fluid to the tip.

In this project, two oils having potential for being developed as cutting fluids were investigated along with a custom-made dispensing system configured to deliver at variable flow rates. The idea is to build a very low-cost MQL system for dispensing cutting fluid. So, a primary analysis was carried out on unblended oils—a straight oil and a synthetic oil (used), with a goal to further develop a better effective cutting fluid along with a delivery system. A simple plain turning operation on a conventional machine tool with a HSS S400 grade single point cutting tool on a MS workpiece is considered for the purpose. Conditions found in a typical small-scale industrial manufacturer, with a job shop-type setup were assumed for this investigation.

## 2 Methodology

### *2.1 Experimental Setup and Design*

**Cutting fluids**. A sample of used or spent engine oil (SEO) was collected from an auto mechanic's garage, at Rs. 10/litre. To reduce the particulate matter, debris from the oil was filtered out using filter papers. Palm oil (PO) was purchased at Rs. 50/litre. The latter was uncontaminated and fresh. Both of these have been used as it is, i.e. without any additives or further processing, in the current investigation. Viscosities of both the fluids were measured. Viscosity variations can be attributed to average molecular sizes, which are controllable in synthetic oils. These synthetic oils, such as the engine oil being used in the current project, do not change viscosity over time of use, otherwise caused by external factors during the conditions of use, which could lead to change in viscosity. This is attributed to a consistent molecular size and structure, resulting in slowing the viscosity change. Table 1 *gives a summary of the basic properties of the two oils being considered in this experimentation.*

**Fluid Dispensing System and its Calibration**. To fabricate a low-cost setup, a hybrid dispensing system was developed basing on two different applications—the drip system typically used in water scarce areas for irrigating crops and the intravenous fluids injection system used for patient care in hospitals. A simple iron stand was fabricated, for holding a saline bottle arrangement, using 12 mm iron rods. The holding arrangement can be varied in height above the ground to increase or decrease the overall head. To build a fluid delivery system, saline bottles, plastic tubes, valves and IV infusion sets were also procured. The valves helped control as well as measure the flow rate of the oil. In order to calibrate the drip system, for two extreme openings of the valve in this setup, the minimum and maximum flow rates

**Table 1** Properties of oils used

| Properties | Typical values | |
|---|---|---|
| | SEO[a] | PO |
| Physical state | Liquid | Liquid |
| Colour | Dark brown | Amber |
| Flash point (°C) | 184 | 280 |
| Pour point (°C) | −10 | 9 |
| Kin. viscosity (cst) (40 °C) | 82.6 | 40.24 |
| Kin. viscosity (cst) (100 °C) | 9.8 | 7.89 |
| pH value | 5.7 | 4.43 |

were identified and fixed. Consequently, the flow rate was matched against various intermediate positions of the valve knob.

**Workpiece and cutting tool**. Mild steel is steel containing small percentage of carbon, strong and tough but not readily tempered. It contains approx. 0.05–0.25% carbon making it malleable and ductile. Being the most common form of steel, its price is relatively low while it provides material properties that are acceptable for many applications. The cutting tool used is S400, a T42 Cobalt-based HSS tool. The tool was ground to its optimum geometry for machining MS workpieces. The tool was configured to 10-7-6-8-15-16-0.4 (2.4), using a tool grinding machine.

**Machine Tool**. A 6 ft. medium duty conventional lathe of 2015 local make is used for the purpose of carrying out the experiments. So, the age of the machine as on the date of conducting the experiment would be approximately 5 years. The feed and speed settings available on the machine were selected. A feed of 0.98 mm/s and 8.0125 mm/s was selected as low and high settings. Similarly, spindle speeds of 315 rpm and 780 rpm were selected.

**Factorial design**. Experiments were based on a fractional factorial design-based Taguchi's orthogonal array. The investigation is designed to test five factors—oil, flow rate, feed rate, doc and RPM—at two levels each, with one of the factors, i.e. oil, being categorical. Flow rate refers to the average oil flow rate from the dispenser. Feed rate is the cutting tool's movement in the longitudinal direction. Depth of cut (doc) refers to the radial feed provided to the cutting tool before machining. RPM refers to the spindle speed. So, an $L_8$ orthogonal array was obtained for 8 runs as shown in Table 2. The corresponding process parameters of the experiment are given in Table 3.

## *2.2 Machining Outputs*

Surface roughness was monitored during this investigation. This response variable was measured offline using a Mitutoyo Surftest SJ-210 instrument. Mean CLA values

**Table 2** Orthogonal array ($L_8$)

| Runs | Factors | | | | |
|---|---|---|---|---|---|
| | A | B | C | D | E |
| 1 | 1 | 1 | 1 | 1 | 1 |
| 2 | 1 | 1 | 1 | 2 | 2 |
| 3 | 1 | 2 | 2 | 1 | 1 |
| 4 | 1 | 2 | 2 | 2 | 2 |
| 5 | 2 | 1 | 2 | 1 | 2 |
| 6 | 2 | 1 | 2 | 2 | 1 |
| 7 | 2 | 2 | 1 | 1 | 2 |
| 8 | 2 | 2 | 1 | 2 | 1 |

**Table 3** Experimental orthogonal array ($L_8$)

| Runs | Factors | | | | |
|---|---|---|---|---|---|
| | Oil | Flow Rate (ml/s) | Feed rate (mm/s) | doc (mm) | $N$ (RPM) |
| 1 | PO | 0.63 | 0.98 | 0.5 | 315 |
| 2 | PO | 0.63 | 0.98 | 1.0 | 780 |
| 3 | PO | 0.90 | 8.0125 | 0.5 | 315 |
| 4 | PO | 0.90 | 8.0125 | 1.0 | 780 |
| 5 | SEO | 0.63 | 8.0125 | 0.5 | 780 |
| 6 | SEO | 0.63 | 8.0125 | 1.0 | 315 |
| 7 | SEO | 0.90 | 0.98 | 0.5 | 780 |
| 8 | SEO | 0.90 | 0.98 | 1.0 | 315 |

of roughness of the turned workpiece were considered after taking four random readings from the surface, using the same cut-off values.

The experimental setup is summarized in Table 4.

## 3 Results and Discussion

An analytical expression for surface roughness obtained on turned components was used to compare the values obtained from experimentation. A relationship was established between roughness and feed rate and cutting tool's nose radius, by Kawalec et al. (1979) and Grzesik (2011). So for larger nose radius, a smoother surface can be obtained. Also, a low feed rate can also result in a better surface finish, for a given nose radius.

$$h = \left[f^2/(8R_n)\right] \times 10^3 \mu\text{m} \quad (1)$$

**Table 4** Experimental setup

| Properties | Typical values |
|---|---|
| Machine | 6 ft medium duty engine Lathe |
| Workpiece material | Mild steel |
| Process parameters | |
| Spindle speed | 315 and 780 |
| Depth of cut (mm) | 0.5 and 1.0 |
| Feed rate (mm/s) | 0.98 and 8.0125 |
| Oil flow rate | 0.63 and 0.90 ml/s |
| Cutting tool | |
| Tool material | Cobalt-based HSS |
| Geometry | 10-7-6-8-15-16-0.4 (2.4) |
| Cutting fluids | Spent engine oil and palm oil |
| Cutting fluid supply | MQL method |

**Table 5** Maximum cutting feed depending upon the nose radius

| Nose radius ($r_n$) (mm) | 0.4 | 0.8 | 1.2 | 1.6 | 2.4 |
|---|---|---|---|---|---|
| Max. cutting feed ($f$) (mm/rev) | 0.25–0.35 | 0.4–0.7 | 0.5 –1.0 | 0.7–1.3 | 1.0–1.8 |

Astakhov (2011) suggested that, in rough turning operations, a large nose radius: (1) increases the strength of the insert; and (2) in inserts permits working with higher cutting feeds. The maximum cutting feed in roughing operations should be selected according to the data shown in Table 5.

Mean surface roughness values measured after every run are used to investigate the performance of both the oils. Statistical analyses of experimental results obtained are carried out using MINITAB 17, a statistical package, which is used for analysis. Images of the experimental setup and machining can be seen in Fig. 3.

## 3.1 *Analysis of S/N and ANOVA*

The statistical analysis has been carried out to find process parameters which can influence the machining quality (surface roughness) and to identify the favourable level of each process parameter, especially, that parameter which is related to the lubri-cooling system. Adopting a Taguchi approach to analysis, the *S/N* ratio values for surface roughness are obtained using the equation shown below. An *S/N* ratio of the smaller-the-better type is used in this paper. So, for optimizing the process in this study, the smaller the *S/N* (dB) ratio, the better is the characteristic. This *S/N* ratio is generally used when the response value is continuous and nonnegative and the desired value is zero.

$$\frac{S}{N} = -10\log_{10}\left[\frac{1}{n}\sum_{k=1}^{n} y_k^2\right] \tag{2}$$

where $n$ is the number of measurements in a trial/row and $y_k$ is the $k$th measured value in a run. It includes the effect of noise as well. Both main effects and interaction effects were observed among some of the factors. Main effects can be observed from all the graphs of the factors on the response variables. The following two-way interactions are found likely:

a. Oil*flow, oil*feed, oil*doc and oil*RPM
b. Flow*feed
c. Doc*rpm.

In case of interaction of oil with other factors, since considerable difference in viscosities exists between the two cutting fluids used, viscosity can be an influential factor in the machining process. So, while viscosity is a factor that can be controlled to influence the response variable, it is important to understand that it is the relative change in viscosity during the process and in the cutting zone which is important.

Main effects on the response variable can be seen (see Fig. 1).

Similarly, interaction effects on the response variable can be seen (see Fig. 2).

As far as two-way interactions are concerned, significant interactions that can be visually observed in the interaction plots are between the oil and its flow, oil and feed, oil and doc, and oil and rpm. While oil and its flow rates can be linked, other

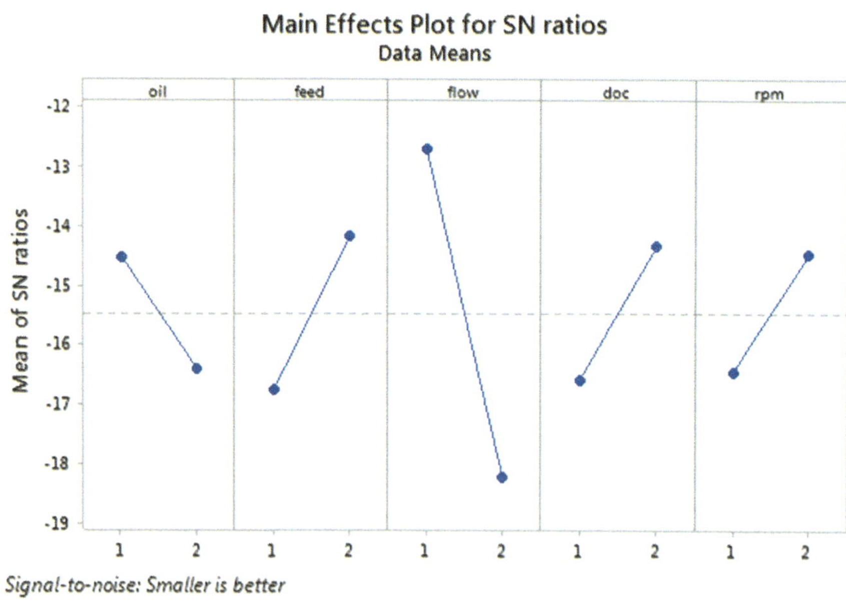


**Fig. 1** Main effects of factors on signal-to-noise ratio of surface roughness

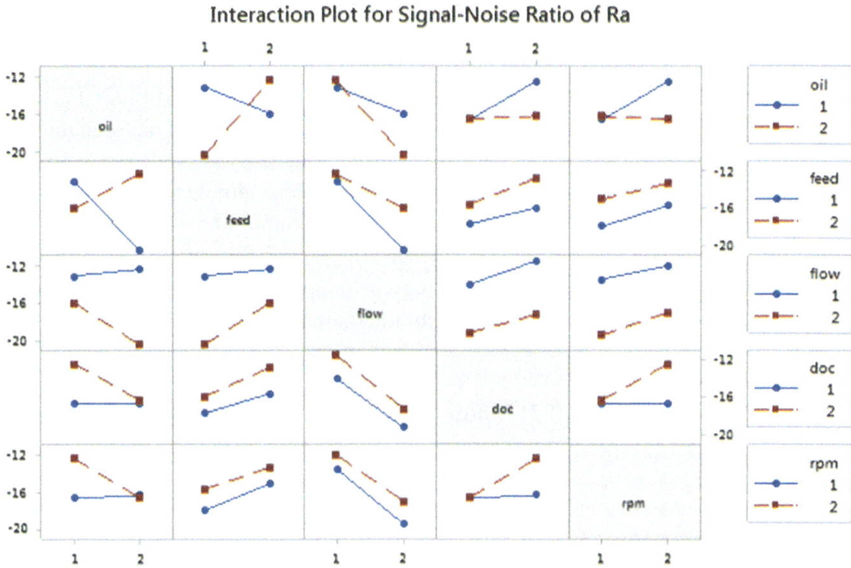


**Fig. 2** Two-way interaction effects on signal-to-noise ratio of surface roughness

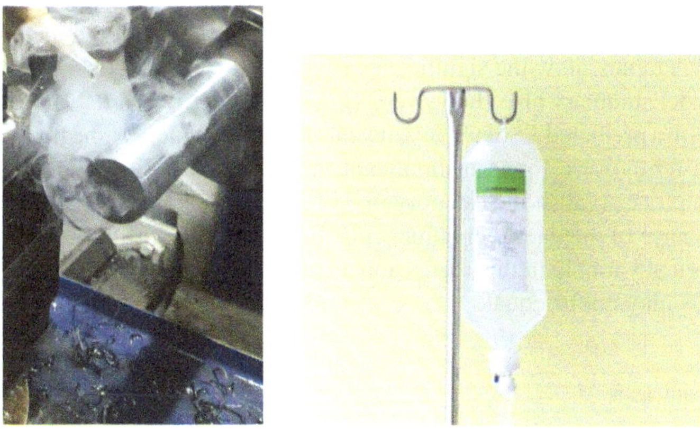

**Fig. 3** Machining under the influence of cutting fluid

linkages can be safely ruled out. Similarly, among the interactions of feed and flow, feed and doc, and feed and rpm, the first interaction seems to be showing a significant effect on the response variable. No arrangement was made to regulate or control one factor via any other factor through any form of mechanism (Fig. 3).

### 3.2 Multiple Regression of Surface Roughness

Multiple regression analysis on the obtained data is done using Minitab 17, after considering various interactions. The interactions detected have been factored into the regression model along with other factors, yielding category-wise (oil-wise) models. The independent variables are flow rates of cutting fluids, spindle speeds, feed rates and depths of cut. The oil type is the categorical variable and is also one of the predictors. The dependent variable is surface roughness. So, two separate models were obtained, after following a stepwise regression to fit a hierarchical model to the observed data. The regression equations obtained are as follows.

PO:

$$-R_a^{-1} = 0.0 - 0.1091\mathbf{feed} - 0.2173\mathbf{flow} + 0.135\mathbf{feed} * \mathbf{flow} - 0.000144\mathbf{rpm} * \mathbf{doc} \tag{1}$$

SEO:

$$-R_a^{-1} = 0.1396 - 0.1091\mathbf{feed} - 0.2173\mathbf{flow} + 0.135\mathbf{feed} * \mathbf{flow} - 0.000144\mathbf{rpm} * \mathbf{doc} \tag{2}$$

The analysis reveals the statistical significance of the regression coefficients of the considered factors and their interactions in Tables 6 and 7. Both the coefficients and ANOVA tables show the significance of the factors considered into the model.

The model summary also shows very high predictability. An $R^2$ value of 99.69% together with a predicted $R^2$ of 98.36% reveals the effectiveness of the fit model. Other interaction terms were found insignificant, in accordance with graphical two-way interaction plots. Against the category of "oil", a $p$-value of 0.006 strongly suggests the significance of this factor in influencing the machining performance. Since "oil" is a categorical variable in this regression analysis, its significance indicates that the level means may not be equal.

**Table 6** Coefficients

| Term | Coef | SE coef | $T$-value | $P$-value |
|---|---|---|---|---|
| Feed | −0.10911 | 0.00993 | −10.98 | 0.002 |
| Flow | −0.2173 | 0.0277 | −7.83 | 0.004 |
| Oil | | | | |
| SEO | 0.1396 | 0.0200 | 6.99 | 0.006 |
| Feed*flow | 0.1350 | 0.0133 | 10.18 | 0.002 |
| rpm*doc | −0.000144 | 0.000028 | −5.16 | 0.014 |

**Table 7** ANOVA of the response

| Term | DF | Adj SS | Adj MS | F-value | *P*-value |
|---|---|---|---|---|---|
| Regression | 5 | 0.300274 | 0.060055 | 195.53 | 0.001 |
| Feed | 1 | 0.037049 | 0.037049 | 120.63 | 0.002 |
| Flow | 1 | 0.018842 | 0.018842 | 61.35 | 0.004 |
| Oil | 1 | 0.014997 | 0.014997 | 48.83 | 0.006 |
| Feed*flow | 1 | 0.031856 | 0.031856 | 103.72 | 0.002 |
| rpm*doc | 1 | 0.008176 | 0.008176 | 26.62 | 0.014 |
| Error | 3 | 0.000921 | 0.000307 | | |
| Total | 8 | 0.301195 | | | |

### 3.3 Sustainability Analysis

A sustainability analysis of the lubri-cooling process was done with reference to three factors—recyclability, cost factor and operator comfort. Recyclability or waste minimization in the current project refers to reusability of the oil for the given machining process. Material saving within the product life cycle may arise from product reuse, recycling arising from waste management practices and/or recycling of energy and materials within manufacturing stages. On this basis and for any individual manufacturing process, the sustainability factor (*R*):

$$SF_R = 1 - [\text{Recycled Material}/\text{Material used initially}] \tag{5}$$

Values of $SF_R$ range between zero and unity. A value of zero is obtained when all available materials have been recycled and unity when there is no recycling. So, lower values of $SF_R$ are preferred. In this project, in case of SEO, its utility was restricted to the first run only; i.e. most of the oil was unusable after the first run of the operation. The quantity of waste was unrecoverable after a single run of the turning operation. However, in case of the palm oil (PO), roughly 35% of the waste oil was recoverable after the entire operation was finished.

Secondly, machining costs are affected by the costs of the cutting fluids consumed in the process. So, in the current case, the sustainability factor (*C*):

$$SF_C = 1 - [\text{Cost of Fluid Used}/\text{Best Alternative Fluid}] \tag{6}$$

Values of $SF_C$ range between zero and unity. A value of one is obtained when the fluid's cost is almost zero or very low compared to the benchmarked fluid and zero when it is as costly as that of the ideal fluid. So, larger values of $SF_C$ are preferred. For dry turning, $SF_C$ would be equal to 1. For this project, HP Koolkut 40 has been considered as the benchmark fluid rated at Rs. 110/litre.

Thirdly, operator comfort can be best expressed as a process that causes the least hazard or hindrance to a human operator in running and controlling the process. If a comfort rating on a scale of 0–9 is provided by a skilled operator, 0 being "Most Uncomfortable" and 9 being "Very Comfortable", the rating provided by the skilled operator is then considered for evaluating the disruption to focusing on the controlling operation and performing effortlessly.

For a rating of 0, the fumes or any emissions from the process would be irritating to the eye and render the person incapable to focus properly on the machine. Also, the emissions can make it difficult to breathe properly and could result in accidents to the operator. In case of 9, a smooth conduct of the machining process can be done, which is identical to a dry machining process at small machining jobs, as far as operator comfort is concerned. So, $SF_O$ is the sustainability factor ($O$):

$$SF_O = 1 - [\text{Comfort Rating}/9] \tag{7}$$

A skilled operator was asked to perform machining with each of the oils and accordingly rate the comfort levels of operating with the particular fluid. Palm oil, in the current case, was releasing excessive fumes whenever the fluid was dropped onto the tool surface, and it came in contact with the hot surface of the tool or the chip. Compared to this, SEO released fairly smaller amounts of fumes, when in contact with the tool or chips. SEO was rated at 7, while PO was rated 2 on the aforementioned scale.

The ratings for each of the above sustainability factors are summarized in Table 8.

Figure 4 shows a comparison between the sustainability scores for both the cutting fluids used. Though the spent engine oil appears to score high on operator comfort and cost factors, its recyclability is low. Moreover, being mineral oil-based, it is also

**Table 8** Sustainability factors

| Sustainability factor | Score type | Symbol | SEO | PO |
|---|---|---|---|---|
| Recyclability | Smaller-the-better | $SF_R$ | 1.00 | 0.65 |
| Cost factor | Larger-the-better | $SF_C$ | 0.91 | 0.54 |
| Operator comfort | Smaller-the-better | $SF_O$ | 0.22 | 0.78 |

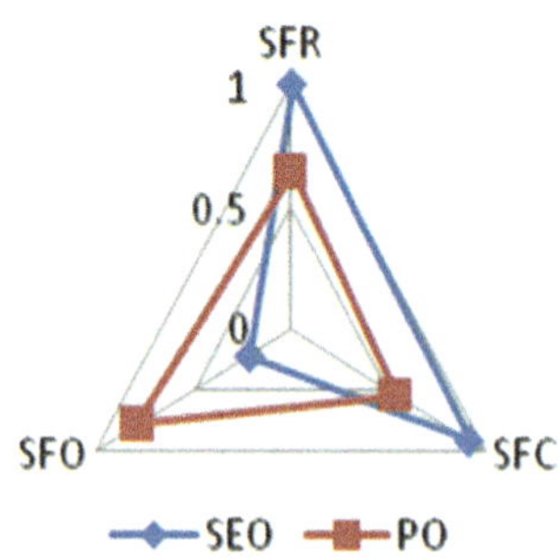


**Fig. 4** Comparison of sustainability factors of both oils

known to be non-biodegradable. On both of these characteristics—recyclability and biodegradability, we know that palm oil can have the highest preference.

### 3.4 *Discussion*

In the present work, an attempt has been made to evaluate the performance of palm oil and spent engine oil as lubricants for turning of mild steel with high-speed steel tool.

- Machining with palm oil as lubricant is found to be better than spent engine oil. Although economically the latter scores high, with respect to recyclability, the vegetable oil was found better. Any attempt to improve recyclability of spent engine oil will result in escalating its cost. Small difference in the response results for palm oil and spent engine oil can be ignored for this very simple reason. So, palm oil is better to use as a lubricant than spent engine oil keeping in view various aspects of health and better for environment as the high hazardous fumes are produced while machining with spent engine oil as lubricant compared to palm oil.
- The optimal and best combination for better surface finish is obtained by using Taguchi methodology are palm oil as lubricant, at moderate cutting speed, low feed rate, more depth of cut and at low flow rate of lubricant. The best combination for low flank wear of HSS tool is spent engine oil as lubricant, low cutting speed, low feed rate, more depth of cut and at high flow rate of lubricant.
- Using ANOVA, individual factor effects of flow and feed factors were found significant. The other factors affecting the response by interactions include feed and flow rate, and spindle speed (rpm) and depth of cut (doc).
- A fairly good control in dispensing the lubricant to the cutting zone has been attained using low-cost components as nozzle and control valves. Also, the ability of oil to flow through the designed mechanism was found to be affected by the ambient temperature. So, a slight heating of the oil enabled it to flow through the dispenser to cutting zone. Also, application of a slight pressure to the fluid might improve the penetrability of the fluid.

## 4 Conclusions and Recommendations

The current investigation has provided a starting point for further study into developing the system's utility on improving its overall sustainability. The following conclusions can be drawn with reference to this study. Since process performance is a key expectation, surface roughness was selected. Since the fluid selection will determine the extent of tool wear and it will have a direct impact on surface finish of the workpiece, it was sufficient monitoring the end effect, though tool wear or life could have been a performance measure also. A significantly better surface finish was observed in case of the vegetable oil used than the recycled synthetic oil, even though the latter are meant to have a large operating range of viscosity due to uniformity in molecular size, as mentioned earlier. Small variations in roughness values were observed across the length of the workpiece. Since temperatures are high, the oil viscosities might have reduced near the cutting zone leading to bad performance of SEO.

Smaller energy consumption can also increase the sustainability of the process. This is possible if lower cutting forces can be applied by reduction of friction using a lubricant in the right quantity. Tool temperature or the chip temperature also could have been monitored to measure the effect of the cutting fluid used. But these also would have an effect on the tool wear, further leading to rougher surfaces on the workpieces. MRR, chip thickness ratio, tool life or wear, power consumption or cutting forces, product attributes like cylindricity, roundness, surface hardness, etc., and oil condition characteristics like particulate matter, pH value, viscosity, flash point, saponification number, oiliness, etc., can be other response variables.

One important issue was the focussed delivery of the cutting fluid to the cutting zone. As the zone was below the chip, only a small portion of the tool was exposed outward and so, the nozzle had to be placed to deliver the fluid drops to this exposed area of the cutting tool. If pressure can be applied to the fluid so that a drop can be forced in between the chip and the tool surface, then the system can be much more effective. Another challenge was the use of the rubber tubing and the plastic nozzle that was getting damaged during the turning process, by the long chips. Rerouting the tubing and replacing the plastic nozzle with a glass one were found to be better options.

The models developed in the above analysis have shown the possibility of accurate prediction using multiple regression. While flood lubrication involves using a surplus quantity of fluid, an MQL will focus on sustainability through a right quantity of fluid. This right quantity of cutting fluid is the optimum quantity that minimizes energy consumption and surface roughness. So, if optimum values of process parameters can be determined, the designed MQL system can be calibrated for different kinds of fluids and different materials and quality outcomes desired.

**Acknowledgements** We acknowledge the support of our host institution, Raghu Engineering College, Visakhapatnam, in extending laboratory facilities and supporting operational expenses of all the related experimentation.

## References

1. Dornfeld DA, Yuan C, Diaz N, Zhang T, Vijayaraghavan A (2013) Green manufacturing: fundamentals and applications. Springer Science + Business Media, New York. https://doi.org/10.1007/978-1-4419-6016-0_9
2. Srikant RR, Rao PN (2017) Use of vegetable-based cutting fluids for sustainable machining. In: Davim J (eds) Sustainable machining. Materials forming, machining and tribology. Springer, Cham. https://doi.org/10.1007/978-3-319-51961-6_2

3. Krishna PV, Srikant RR, Rao DN (2010) Experimental investigation on the performance of nanoboric acid suspensions in SAE-40 and coconut oil during turning of AISI 1040 steel. Int J Mach Tools Manuf 50:911–916
4. Rao SN, Satyanarayana B (2011) Experimental estimation of tool wear and cutting temperatures in MQL using cutting fluids with CNT inclusion. Int J Eng Sci Technol 3(4):928–931
5. Khandekar S, Sankar MR, Agnihotri V, Ramkumar J (2012) Nano-cutting fluid for enhancement of metal cutting performance. Mater Manuf Processes 27(9):963–967
6. Shah P, Gadkari A, Sharma A, Shokrani A, Khanna N (2021) Comparison of machining performance under MQL and ultra-high voltage EMQL conditions based on tribological properties. Tribol Int 153, 106595. ISSN 0301-679X. https://doi.org/10.1016/j.triboint.2020.106595
7. Venkata Vishnu A, Jamaleswara Kumar P (2018) Investigations on high speed machining of EN-353 steel alloy under different machining environments. In: IOP conference series: materials science and engineering 330
8. Arul K, Senthil Kumar VS (2020) Magnetorheological based minimum quantity lubrication (MR-MQL) with additive n-CuO. Mater Manuf Processes 35(4):405–414. https://doi.org/10.1080/10426914.2020.1732410
9. Odusote JK, Kolawole (2015) Assessment of suitability of vegetable-based oils as cutting fluids in drilling operation. Ann Facul Eng Hunedoara-Int J Eng XIII:169–174
10. Kamata Y, Obikawa T (2007) High speed MQL finish turning of Inconel 718 with different coated tools. J Mater Process Technol 192–193:281–286. https://doi.org/10.1016/j.jmatprotec.2007.04.052
11. Sen B, Mia M, Krolczyk GM et al (2019) Eco-friendly cutting fluids in minimum quantity lubrication assisted machining: a review on the perception of sustainable manufacturing. Int J Precis Eng Manuf-Green Tech. https://doi.org/10.1007/s40684-019-00158-6
12. Astakhov VP (2008) Ecological machining: near-dry machining. In: Machining, pp 195–223. Springer, London
13. Wang J, Zhang J (2005) On formation and breakup of boundary lubricating layer. Run Hua Yu Mi Feng/Lubric Eng 6:001
14. Vishnu A, Jamaleswara Kumar P, Ramana V, Maparla. (2018) Comparison among dry, flooded and MQL conditions in machining of EN 353 steel alloys-an experimental investigation. Mater Today Proc 5:24954–24962. https://doi.org/10.1016/j.matpr.2018.10.296
15. Venkata Vishnu A, Jamaleswara Kumar P (2019) Experimental investigations in turning of steel alloy under different lubrication conditions. In: Pujari S, Srikiran S, Subramonian S (eds) Recent advances in material sciences. Lecture notes on multidisciplinary industrial engineering. Springer, Singapore. https://doi.org/10.1007/978-981-13-7643-6_21
16. Vishnu A (2017) Experimental investigation on effect of machining parameters of EN353 alloy steel using nano fluids 2:2017–2395
17. Venkata Vishnu A, Venkata Ramana M, Tilak KBG (2017) Experimental investigations of process parameters influence on surface roughness in turning of EN-353 alloy steel under different machining environments. Mater Today Proc 5(2), Part 1:4192–4200. ISSN 2214-7853, https://doi.org/10.1016/j.matpr.2017.11.682
18. Matta AK, Kodali SP, Ivvala J, Kumar PJ (2018) Metal prototyping the future of automobile industry: a review. Mater Today Proc 5(9) Part 3:17597–17601
19. Krishna Reddy KSV, Kabra N, Kunchum U, Vijayakumar T (2014) Experimental investigation on usage of palm oil as a lubricant to substitute mineral oil in CI engines. Chin J Eng 2014, Article ID 643521, 5 pages. https://doi.org/10.1155/2014/643521
20. Nwadinobi CP, Okoli JU (2018) Development of cutting fluid from spent engine oil. Nigerian J Technol (NIJOTECH) 37(4):950–956. https://doi.org/10.4314/njt.v37i4.13
21. Ramasamy D, Ramachandran K, Yeesvaran B, Kadirgama K, Anwar Che Ghani S, Anamalai K (2017) State of art of cooling method for dry machining. In: AiGEV 2016, MATEC web of conferences 90, 01015. https://doi.org/10.1051/matecconf/20179001015
22. Culaba AB, Purvis MRI (1999) A methodology for the life cycle and sustainability analysis of manufacturing processes. J Cleaner Prod 7:435–445

23. Hassan AB, Abolarin MS, Nasir A, Ratchel U (2006) Investigation on the use of palm olein as lubrication oil. Leonardo Electron J Pract Technol 5(8):1–8
24. Brockhoff T, Walter A (1998) Fluid minimization in cutting and grinding. In: Abrasives Magazine, pp 38–42
25. Gelfi AM, Giardini C (1995) Minimum quantity lubrication in turning. Wear 260
26. Kolawole S, Odusote J (2013) Performance evaluation of vegetable oil-based cutting fluids in mild steel machining. Chem Mater Res 3(9). ISSN 2224-3224 (Print) ISSN 2225-0956 (Online)
27. Ashtakov VP (2011) Modern machining technology—a practical guide. Woodhead Publishing, pp 1–78. https://doi.org/10.1533/9780857094940.1
28. Rifat M, Habibor RM, Das D (2017) A review on application of nanofluid MQL in machining. AIP Conf Proc 1919:020015. https://doi.org/10.1063/1.5018533
29. Roy S, Ghosh A (2013) High speed turning of AISI 4140 steel using nanofluid through twin jet SQL system. In: Proceedings of the ASME 2013 international manufacturing science and engineering conference collocated with the 41st North American manufacturing research conference, vol 2: systems; micro and nano technologies; sustainable manufacturing. Madison, Wisconsin, USA. June 10–14, 2013. V002T04A002. ASME. https://doi.org/10.1115/MSEC2013-1067
30. Sharma AK, Tiwari AK, Dixit AR (2015) Progress of nanofluid application in machining: a review. Mater Manuf Processes 30(7):813–828. https://doi.org/10.1080/10426914.2014.973583
31. Posinasetti N (2013) Sustainable manufacturing—principles, applications and directions. In: 28th national convention of production engineers, institution of engineers India, At: MNIT, Jaipur, vol 28
32. Srikanth M, Sai Ravindra P, Harsha Vardhan DV, Sai D, Ramamohan Rao G (2020) Parametric analysis on plain turning operation under the influence of minimally dispersed sustainable cutting fluids, B. Tech final year project report, 2016–2020, Raghu Engineering College, Visakhapatnam

# The Effect of Drilling Parameters on the Hole Quality of Hybrid Fiber-Reinforced Epoxy Composite

**V. Santhanam, S. Sendhilkumar, N. Venkateshwaran, and M. Chandrasekaran**

**Abstract** In this work, hybrid fiber-reinforced epoxy composite had been developed by using woven banana fiber and woven glass fiber as reinforcements. Laminated composite sample was developed by using hand layup method. Glass fiber was kept as top layer, and banana fiber was kept as bottom layer with the arrangement of G-B-G-B. Drilling is one of the most common and critical machining operations during the assembly operation in manufacturing. In this work, the effect of drilling parameters on the quality of the hole and delamination was studied. Three different drill speeds and feed rates were used in this study. The machining was done by using a twist drill of diameter 8 mm, and the drilling was performed at the speeds of 700, 1400 and 2100 rpm. The feed rate was kept at 0.05, 0.12 and 0.20 mm/rev. The drilled holes were scanned in a high-resolution scanner, and digital image processing was further used to compute the hole dimensions. The details of the pixel values at the entry and exit were measured digitally, and subsequently, the pixel values were used to determine the extent of damage occurred due to the drilling operation. A two-dimensional array of hole dimension with respect to the feed rate and speed was obtained from the experimental results. Finally, ANOVA technique was used to determine the most influencing drilling parameter to get better hole quality. The experimental results indicated that the damage due to machining was greater at the exit than the damage at the entry. Also, ANOVA technique showed that the drilled hole quality depends on the drilling speed and feed rate, but the feed rate is the most significant parameter than the cutting speed.

**Keywords** Composite drilling · Hybrid fiber-reinforced composite · Banana fiber · Glass fiber · ANOVA technique

---

V. Santhanam (✉) · N. Venkateshwaran
Rajalakshmi Engineering College, Chennai, India

S. Sendhilkumar
Akshaya College of Engineering and Technology, Coimbatore, India

M. Chandrasekaran
Vels Institute of Science Technology and Advanced Studies, Chennai, India

 
P. Srinivasa Pai and V. Krishnaraj (eds.), *Sustainable Machining Strategies for Better Performance*, Lecture Notes in Mechanical Engineering,
https://doi.org/10.1007/978-981-16-2278-6_9

# 1 Introduction

Nowadays, the demand for the multifunctional materials is becoming prime importance in automotive applications. The main task is to develop high-performance, low-cost and low-density materials with better strength and modulus along with good thermal characteristics. This diverse and contradictory requirement for the automotive materials has brought the very complex problem of developing new types of polymer composites materials. Polymer composite materials had been increasingly used in several applications due to their advantages such as low weight to strength ratio, better wear properties, ease of manufacturing, and the properties can be easily tailor-made as per the requirements of the application. Due to the increasing environmental concerns and carbon emission regulations, recently the focus is on the use of natural fibers as a successful reinforcement in place of synthetic fibers such as glass fiber, carbon fiber and Kevlar fiber. But the natural fibers are hydrophilic, and the mechanical properties of the plant fibers are inferior to that of the synthetic fibers. Hence, attempts are being made to develop the hybrid fiber-reinforced polymer composites where both the natural fiber and synthetic fiber are used as the reinforcement to get the advantage of both the fibers. Hybridization [1] is one of the effective means of achieving desired properties of laminated composite by combining two or more types of fiber or matrix. Also, the hybridization suppresses the weakness of the constituents due to the addition of better material in it, and thereby, the compensation of properties also occurs [2]. By developing such versatile materials, the machining of the material becomes difficult to accommodate this development. Of the various available secondary process of machining the composite materials, drilling is most widely used during the assembly process. The quality of drilled holes in composite materials involves the parameters like feed rate, speed, drill wear and drill geometry. The machinability parameters include torque, thrust force, residual strength, surface roughness and damages associated.

Problems such as delamination, hole quality and excessive tool wear are frequently encountered during the drilling process of the polymer composite materials. The delamination in composite material is a significant factor as it affects the structural strength, poor assembly tolerance and micro-cracking of the components during the usage [3]. The effect of drilling parameters on the tool wear and delamination of glass fiber composite was investigated by Prakash et al. [4] with the help of vibration signal. The time domain and frequency domain analysis was carried out to determine the optimum drilling parameters, and from the experimental results, it was found that the frequency domain analysis predicted better results than the time domain analysis. The results showed that the machining speed of 951 rpm and feed rate of 1 mm/s produced the lowest delamination factor. Sumesh et al. [5] developed composite materials based on pineapple and flax fibers, and cellulose microfiber was used as filler material. Research on machinability of hybrid composite was carried out, and it was shown that the addition of filler up to 3% resulted in better machinability. Various kinds of fillers such as carbon black, coconut shell powder, fly ash and nano alumina were used as secondary reinforcement. The effect of boron nitride particle on the machinability of carbon fiber epoxy composite was carried out by Burak Kaybal et al. [6], and it was observed that the delamination factor decreased with the addition of filler. The majority of studies agree that there is a good correlation between the thrust force and the delamination. Bosco et al. [7] analyzed the influence of machining parameters on thrust force in drilling GFRP-armor steel sandwich composite and showed that the thrust force required is higher for the bottom panel due to the heat generation after drilling the armor steel composite. Rajamurugan et al. [8] studied the drilling performance of the aluminum foil reinforced sisal/flax epoxy composite. The experimental results revealed that the hole pitch and spindle speed greatly affected the circularity of drilled hole. Anand et al. [9] optimized the drilling parameters using Taguchi orthogonal array and grey relational algorithm of Ni-P coated glass fiber and aluminum oxide nano wire hybrid composite. Bukhari et al. [10] studied the influence of drilling parameters of hybrid composite and the influence of stacking sequence on the drilled hole quality. Tran et al. [11] investigated the thrust force and particle generation of drilling the miscanthus fiber with biochar and polypropylene composite. The experiments showed that the feed rate and drill tool diameter had significant effect on the fine and ultra-fine particle emission. Jaganatha et al. [12] analyzed the thrust force and delamination of glass fiber–carbon fiber epoxy composite. From the literature, it was found that the drilling of FRP composites has several undesirable effects such as spalling [13], edge chipping [14], macroscopic pitting [15], fiber breakage, debonding [16], fiber pullout [17], crack formation [18], thermal damage [19], fuzzing [20], stress concentration [21] and matrix crazing along with delamination [22]. From the literature, it was evident that most of the studies on the machining of composite materials were carried out to determine the drilling parameters and drilled hole quality. The effect of fiber loading and drilling parameters on the thrust force, delamination were studied by several researchers. The effect of drilling parameters on the hybrid fiber-reinforced composite and the measurement of hole quality using machine vision technique was not widely reported. Hence, in this work, the effect of drilling parameters on the drilled hole quality of the hybrid fiber-reinforced epoxy was analyzed using machine vision technique.

## 2 Materials and Methods

Banana fiber was used as the natural reinforcement in this work; processed banana fiber was purchased from Rope International Pvt Ltd., Chennai, India. The properties of banana fiber are better than most of the cellulosic fiber available in nature. It has an approximate tensile strength of 600 MPa and modulus of 17.85 GPa. The density of the fiber was estimated as 1.35 gm/cm$^3$. The banana fiber was dried under the sun to remove any residual water content from the fiber. E-Glass fiber was used as synthetic fiber reinforcement in this work, which is procured from Sakthi Fibers, Chennai. It has a density of 2.55 gm/cm$^3$, and the tensile strength is 2400 MPa. The glass fiber was mixed with the banana fiber in equal amounts to prepare the hybrid reinforcement. Epoxy resin was used as the matrix material in this work, which is procured from Sakthi fibers, Chennai. Hybrid fiber composites were fabricated by using hand layup method. A wooden mold of 300 mm × 300 mm was used for fabricating the specimens. Woven fiber arrangement had been used for the purpose. Composites were prepared by using the G-B-G-B arrangement of the woven mats. Figures 1 and 2 show the woven mats of glass fiber and banana fiber used in this study and the composite specimen used for machining operation.

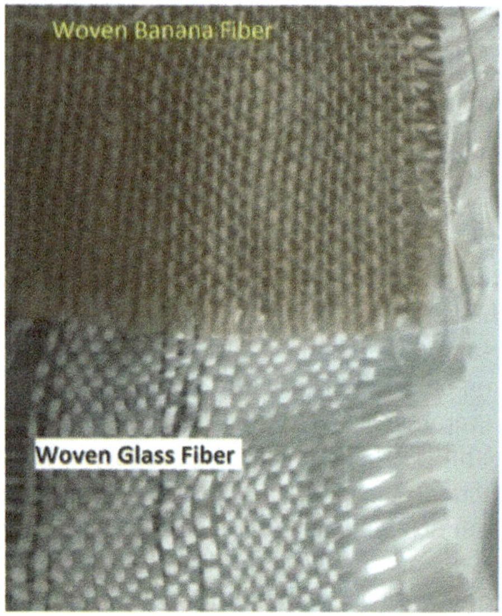


**Fig. 1** Woven banana fiber and woven glass fiber mats

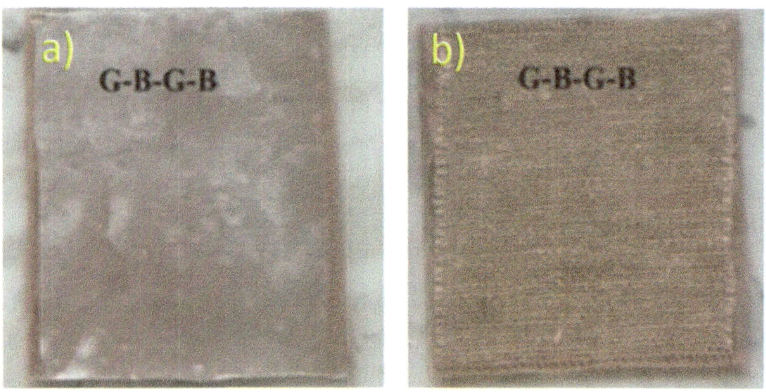


**Fig. 2** Top side (**a**) and bottom side (**b**) of the banana–glass fiber-reinforced hybrid composite specimen

## 2.1 *Experimental*

The drilling operation was performed by using MTAB-DENFORD CNC machine which is shown in Fig. 3. A Single point HSS standard twist drill of 8 mm diameter was used as the drilling tool. The drilling operation was performed at 700, 1400 and 2100 rpm speeds. Considering the diameter of the drill tool, the tangential velocity of the operation was kept between 20 and 60 m/min. The feed rate was varied as 0.05, 0.12 and 0.20 mm/rev. The specimen size of 30 mm × 30 mm was used for the study. The quality of the drilled holes was assessed by using the delamination factor. Machine vision technique was used to measure the hole quality which is shown in Fig. 4. The ratio between the nominal diameter ($D$) of the hole to the largest diameter ($D_{max}$) of the delamination zone will give delamination factor ($F_d$) as shown in Fig. 5.

# 3 Results and Discussion

The drilled hole size at the entry and exit was determined by measuring the pixel value using the machine vision technique. Tables 1 and 2 present the pixel values at the entry (top side) and at the exit (bottom side) of the drilled holes. It can be observed from the tables that the hole size is marginally higher at high feed rate, indicating

**Fig. 3** Drilling machine used in this study

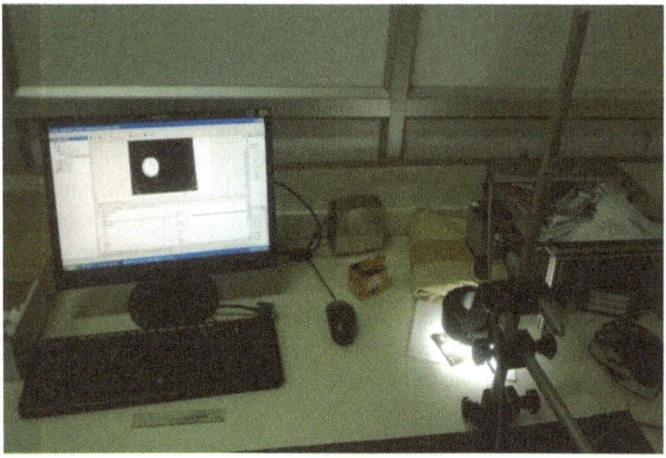

**Fig. 4** Hole quality analysis using machine vision setup

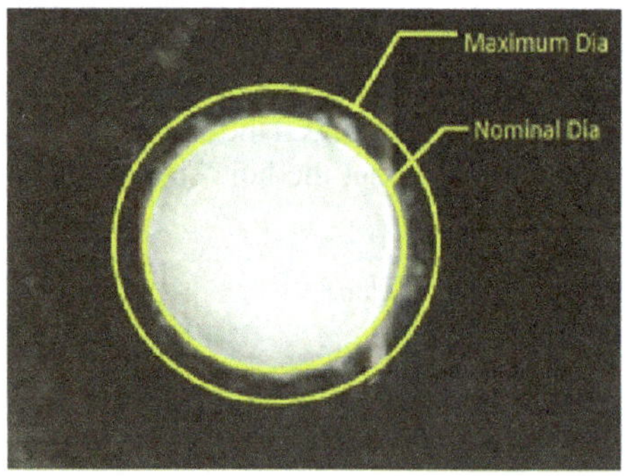


**Fig. 5** Digital image processing of drilled hole

**Table 1** Pixel values and equivalent hole dimension at entry

| Top side | | | | | | | | | | | | |
|---|---|---|---|---|---|---|---|---|---|---|---|---|
| Feed (mm/rev) | 0.20 | | | | 0.12 | | | | 0.05 | | | |
| Speed (RPM) | Pixel | mm | Pixel | mm | Pixel | mm | Pixel | mm | Pixel | mm | Pixel | mm |
| 2100 | 589 | 10.91 | 595 | 11.02 | 560 | 10.37 | 554 | 10.26 | 473 | 8.76 | 468 | 8.67 |
| 1400 | 599 | 11.09 | 593 | 10.98 | 542 | 10.03 | 547 | 10.13 | 465 | 8.6 | 460 | 8.52 |
| 700 | 597 | 11.06 | 603 | 11.18 | 538 | 9.97 | 533 | 9.87 | 482 | 8.92 | 477 | 8.84 |

**Table 2** Pixel values and equivalent hole dimension at exit

| Bottom side | | | | | | | | | | | | |
|---|---|---|---|---|---|---|---|---|---|---|---|---|
| Feed (mm/rev) | 0.20 | | | | 0.12 | | | | 0.05 | | | |
| Speed (RPM) | Pixel | mm | Pixel | mm | Pixel | mm | Pixel | mm | Pixel | mm | Pixel | mm |
| 700 | 681 | 12.62 | 675 | 12.5 | 610 | 11.3 | 616 | 11.42 | 547 | 10.13 | 542 | 10.03 |
| 1400 | 656 | 12.14 | 649 | 12.02 | 612 | 11.33 | 606 | 11.22 | 533 | 9.87 | 538 | 9.97 |
| 2100 | 643 | 11.9 | 636 | 11.78 | 593 | 10.99 | 599 | 11.1 | 530 | 9.81 | 524 | 9.71 |

more damage in the drilled hole. The average values of the delamination factor are presented in Tables 3 and 4, respectively, for at the entry and exit of the drilled holes. It is noted from the tables that the maximum value of delamination factor was

**Table 3** Entry delamination factor for hybrid fiber composite sample

| Fiber arrangement in the composite | Speed (rpm) | Feed rate (mm/rev) | Delamination factor at entry |
|---|---|---|---|
| G-B-G-B | 2100 | 0.20 | 1.37 |
| | 2100 | 0.12 | 1.29 |
| | 2100 | 0.05 | 1.09 |
| | 1400 | 0.20 | 1.38 |
| | 1400 | 0.12 | 1.26 |
| | 1400 | 0.05 | 1.07 |
| | 700 | 0.20 | 1.39 |
| | 700 | 0.12 | 1.24 |
| | 700 | 0.05 | 1.11 |

**Table 4** Exit delamination factor for hybrid fiber composite sample

| Fiber arrangement in the composite | Speed (rpm) | Feed rate (mm/rev) | Delamination factor at exit |
|---|---|---|---|
| G-B-G-B | 2100 | 0.20 | 1.48 |
| | 2100 | 0.12 | 1.38 |
| | 2100 | 0.05 | 1.22 |
| | 1400 | 0.20 | 1.51 |
| | 1400 | 0.12 | 1.41 |
| | 1400 | 0.05 | 1.24 |
| | 700 | 0.20 | 1.57 |
| | 700 | 0.12 | 1.42 |
| | 700 | 0.05 | 1.26 |

observed for the feed rate of 0.20 mm/rev. A maximum value of delamination factor was obtained as 1.39 for entry (top side) and 1.57 at exit (bottom side). The top layer is subjected to compression at the entry of the drilling tool, and hence, less damage is observed; whereas, the bottom layer is subjected to tensile stress during the exit of the drilling tool, and hence, the damage and delamination can easily occur at the bottom. It is because of this reason; woven banana mat is placed at the bottom of the specimen. Since banana fiber has relatively lesser tensile strength than glass fiber mat, the delamination due to the damage of the banana fiber mat is lesser than that of the glass fiber mat.

The effect of drilling speed and feed rate on the entry and exit hole quality is presented in Figs. 6 and 7, respectively. It was observed from the figures that the delamination factor was largely affected by the feed rate rather than the drilling speed.

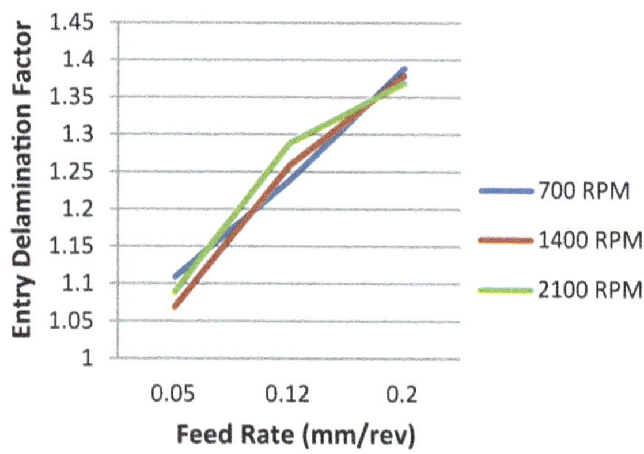


**Fig. 6** Effect of feed rate and drilling speed on entry delamination

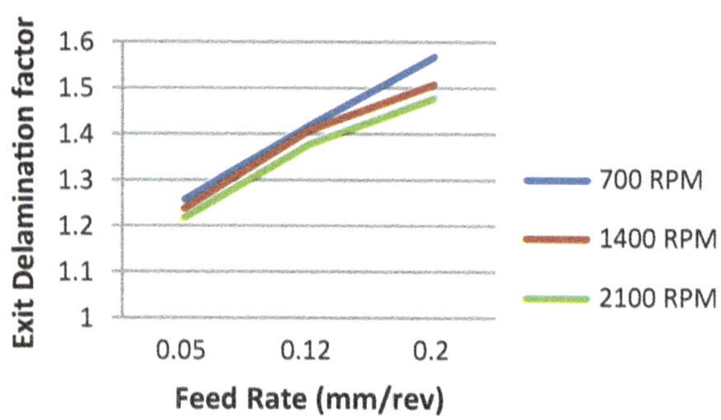


**Fig. 7** Effect of feed rate and drilling speed on exit delamination

## 3.1 ANOVA Analysis

Analysis of variance technique was used to determine the most influential machining parameter for the drilled hole quality. The analysis was carried out by considering two factors (drilling speed and feed rate) with three variables for each factor. Minitab 6.0 software was used for the purpose of analysis. The ANOVA results are presented in Tables 5 and 6, respectively, for the entry and exit delamination. It was noted from the table that the '$p$' value is less than 0.05 at 95% confidence level for the feed rate, indicating that the feed rate has higher significant effect on the hole quality than the other parameters. It was also observed that the value of '$p$' is less than 0.05 for the drill speed at the exit indicating that both the speed and feed rate influence the exit delamination factor.

Regression equations were developed based on the experimental values of entry and exit delamination. These equations can be further used to predict the delamination factor for any value of feed rate and drill speed. Tables 7 and 8 show the values predicted by the regression equation and the values obtained experimentally for the entry and exit delamination. The error between the actual values and predicted values is also presented in Tables 7 and 8 for entry and exit delamination, respectively. The regression models showed an error of 3.96% and 1.97%, respectively, for the entry and

**Table 5** ANOVA for entry delamination for woven fiber composite

| Source | DF | Adj SS | Adj MS | *F*-value | *P*-value |
|---|---|---|---|---|---|
| Speed (rpm) | 2 | 0.0002896 | 0.0001446 | 0.291 | 0.7611 |
| Feed rate (mm/rev) | 2 | 0.1277565 | 0.0638785 | 129.191 | 0.0000 |
| Error | 4 | 0.0019785 | 0.0004934 | | |
| Total | 8 | 0.1300226 | | | |

*Model summary*

| S | R-sq | R-sq(adj) | R-sq(pred) |
|---|---|---|---|
| 0.02223613 | 98.4 | 96.9% | 92.3% |

**Table 6** ANOVA for exit delamination for woven fiber composite

| Source | DF | Adj SS | Adj MS | *F*-value | *P*-value |
|---|---|---|---|---|---|
| Speed (rpm) | 2 | 0.0048221 | 0.0024111 | 9.232 | 0.0324 |
| Feed rate (mm/rev) | 2 | 0.1186893 | 0.0593442 | 227.283 | 0.0000 |
| Error | 4 | 0.0010442 | 0.0002613 | | |
| Total | 8 | 0.1245564 | | | |

*Model summary*

| S | R-sq | R-sq(adj) | R-sq(pred) |
|---|---|---|---|
| 0.0161587 | 99.2% | 98.3% | 95.7% |

**Table 7** Comparison between experimental and regression model results at entry

| Speed (rpm) | Feed rate (mm/rev) | Regression model $F_d$ (entry) | Experimental $F_d$ (entry) | Error (%) |
|---|---|---|---|---|
| 2100 | 0.20 | 1.39 | 1.37 | −1.66 |
| 2100 | 0.12 | 1.24 | 1.29 | 3.96 |
| 2100 | 0.05 | 1.10 | 1.09 | −1.32 |
| 1400 | 0.20 | 1.39 | 1.38 | −0.82 |
| 1400 | 0.12 | 1.24 | 1.26 | 1.78 |
| 1400 | 0.05 | 1.10 | 1.07 | −3.08 |
| 700 | 0.20 | 1.39 | 1.39 | 0.00 |
| 700 | 0.12 | 1.24 | 1.24 | 0.00 |
| 700 | 0.05 | 1.10 | 1.11 | 0.76 |

**Table 8** Comparison between experimental and regression model results at exit

| Speed (rpm) | Feed rate (mm/rev) | Regression model $F_d$ (exit) | Experimental $F_d$ (exit) | Error (%) |
|---|---|---|---|---|
| 2100 | 0.20 | 1.50 | 1.48 | −1.54 |
| 2100 | 0.12 | 1.35 | 1.38 | 1.87 |
| 2100 | 0.05 | 1.22 | 1.22 | 0.00 |
| 1400 | 0.20 | 1.53 | 1.51 | −1.38 |
| 1400 | 0.12 | 1.38 | 1.41 | 1.97 |
| 1400 | 0.05 | 1.25 | 1.24 | −0.99 |
| 700 | 0.20 | 1.56 | 1.57 | 0.71 |
| 700 | 0.12 | 1.41 | 1.42 | 0.69 |
| 700 | 0.05 | 1.28 | 1.26 | −1.61 |

exit delamination of the drilled holes. The negative error means that the delamination factor predicted by the regression model is higher than the experimental value.

This shows that the regression models given by Eqs. (1) and (2) had been able to predict the delamination factor with reasonable degree of accuracy.

Entry delamination = 1.00411 + 0.0000021*Speed (rpm) + 1.9222*Feed rate (mm/rev) (1)

Exit delamination = 1.21539 – 0.0000402*Speed (rpm) + 1.8569*Feed rate (mm/rev) (2)

## 4 Conclusion

Banana–glass hybrid fiber-reinforced composites were developed, and the drilling behavior of the composite was studied using the machine vision technique in this experiment. The effect of drilling speed and feed rate was analyzed to assess the most influencing parameter. ANOVA technique had used for the analysis. The following conclusions were drawn from the experiment.

- The delamination factor increases with both the cutting parameters (velocity and feed rate), which means that the composite damage is bigger for higher cutting speed and higher feed.
- Feed rate had been the most influencing parameter than the cutting speed which is validated by ANOVA technique.
- The delamination factor predicted by regression models was in good correlation with the experimental ones. The maximum error in prediction of entry delamination is 3.96%, and exit delamination is 1.97%.

## References

1. Gupta MK, Srivastava RK (2016) A review on characterization of hybrid fiber reinforced polymer composite. Am J Polym Sci Eng 4(1):1–7
2. Santhanam V, Chandrasekaran M, Venkateshwaran N (2014) Effect of fiber parameters on the mechanical properties of banana-glass fiber hybrid composites. Appl Mech Mater 592–594:202–205
3. Khashaba UA (2013) Drilling of polymer matrix composites: a review. J Compos Mater 47(15):1817–1832
4. Prakash M, Dileep Aditya Dhar PVS (2018) Investigation on the effect of drilling parameters on the tool wear and delamination of glass fiber-reinforced polymer composite using vibration signal analysis. J Compos Mater 52(12):1641–1648
5. Sumesh KR, Kanthavel K, Kavimani V (2020) Machinability of hybrid natural fiber reinforced composites with cellulose micro filler incorporation. J Compos Mater 0021998320918020
6. Burak Kaybal H, Unuvar A, Kaynak Y, Avcı A (2020) Evaluation of boron nitride nanoparticles on delamination in drilling carbon fiber epoxy nanocomposite materials. J Compos Mater 54(2):215–227
7. Bosco MAJ, Palanikumar K, Prasad BD, Velayudham A (2015) Analysis on influence of machining parameters on thrust force in drilling GFRP-armor steel sandwich composites. J Compos Mater 49(13):1539–1551
8. Rajamurugan G, Sanjay AP, Krishnasamy P, Muralidharan B, Jain S (2020) Drilling and mechanical performance analysis on flax-sisal hybrid composite embedded with perforated aluminum foil. J Reinf Plast Compos 39(23–24):902–917
9. Anand G, Alagumurthi N, Elansezhian R, Palanikumar K, Venkateshwaran N (2018) Investigation of drilling parameters on hybrid polymer composites using grey relational analysis, regression, fuzzy logic, and ANN models. J Brazil Soc Mech Sci Eng 40(4):214
10. Bukhari SM, Kandasamy J, Hussain MM (2017) Investigations on drilling process parameters of hybrid composites with different stacking sequence. Mater Today Proc 4(2):2184–2193
11. Tran DS, Songmene V, Ngo AD, Kouam J, Rodriguez-Uribe A, Misra M, Mohanty AK (2020) Experimental investigation on machinability of polypropylene reinforced with miscanthus fibers and biochar. Materials 13(5):1181
12. Jagannatha TD, Bhaskar HB, Sheriff ZA, Irfan G (2019) Optimization of machining parameters of hybrid fiber reinforced polymer composites using design of experiments. In AIP conference proceedings 2057(1), 020008
13. Liu D, Tang Y, Cong WL (2012) A review of mechanical drilling for composite laminates. Compos Struct 94(4):1265–1279

14. Won MS, Dharan CKH (2002) Chisel edge and pilot hole effects in drilling composite laminates. J Manuf Sci Eng 124(2):242–247
15. Zhang LB, Wang LJ, Liu XY (2001) A mechanical model for predicting critical thrust forces in drilling composite laminates. Proc Inst Mech Eng Part B: J Eng Manuf 215(2):135–146
16. Arul SVLMS, Vijayaraghavan L, Malhotra SK, Krishnamurthy R (2006) The effect of vibratory drilling on hole quality in polymeric composites. Int J Mach Tools Manuf 46(3–4):252–259
17. Zhang HJ, Chen WY, Chen DC, Zhang LC (2001) Assessment of the exit defects in carbon fiber-reinforced plastic plates caused by drilling. Key Eng Mater 196:43–52
18. Bhatnagar N, Singh I, Nayak D (2004) Damage investigation in drilling of glass fiber reinforced plastic composite laminates. Mater Manuf Processes 19(6):995–1007
19. Ramirez C, Poulachon G, Rossi F, M'Saoubi R (2014) Tool wear monitoring and hole surface quality during CFRP drilling. Procedia CIRP 13:163–168
20. Abrão AM, Faria PE, Rubio JC, Reis P, Davim JP (2007) Drilling of fiber reinforced plastics: A review. J Mater Process Technol 186(1–3):1–7
21. Rajakumar IPT, Hariharan P, Srikanth I (2013) A study on monitoring the drilling of polymeric nanocomposite laminates using acoustic emission. J Compos Mater 47(14):1773–1784
22. Palanikumar K, Srinivasan T, Rajagopal K, Latha B (2016) Thrust force analysis in drilling glass fiber reinforced/polypropylene (GFR/PP) composites. Mater Manuf Processes 31(5):581–586

# Effect of Profile Geometry and Cutting Speed Override Parameter on Profiling Speed During Tapering Using Wire Electric Discharge Machining

**I. V. Manoj and S. Narendranath**

**Abstract** The oblique/tapered form of precise components has many applications like dies, nozzles, inserts, cutting tools, and other components. Wire electric discharge machining is an erosion technique that helps in precise machining of hard materials. In the present study, basic profiles were machined using a novel slant-type fixture to achieve a slant surface on Hastelloy X. The shapes like triangle, square, and circle are machined at different slant angles, namely 0° and 30°. This paper aims to study the effects of the cutting speed override parameter and profile geometry on profiling speed of machined profiles. The basic shapes of 1, 3, and 5 mm sides were machined. The cutting speed override parameter affected the most on profiling speed in both the angles irrespective of profiles. The profile geometry also affects the profiling speed although the machining parameters were maintained constant.

**Keywords** Slant-type taper profiling · Cutting speed override · Profile geometry

## 1 Introduction

Wire electric discharge machining (WEDM) works on an electro-thermal mechanism which melts the hard conductive workpiece by sparks generated by the electrode. It proves to be the most beneficent machining process, as conventional machining of hard materials may produce many flaws on tool and workpiece [1, 2]. Many materials like ceramics, nanoceramics, composites, titanium-based alloys, nickel-based superalloys, etc., can be machined with precision using EDM [3–6]. Many studies have been carried out on output parameters like cutting speed, surface roughness, kerf, etc. Mouralova et al. [7] have investigated the effects of machining parameters on cutting speed and surface quality. The optimum cutting speed was experimentally confirmed for efficient machining to get a good surface quality. He et al. [8] have investigated machining speed and surface roughness in machining of wire electrical discharge machining (WEDM) of 2D C/SiC composite. Dey and Pandey [9] have

I. V. Manoj (✉) · S. Narendranath
Department of Mechanical Engineering, NITK, Surathkal, Mangalore 575025, India



P. Srinivasa Pai and V. Krishnaraj (eds.), *Sustainable Machining Strategies for Better Performance*, Lecture Notes in Mechanical Engineering,
https://doi.org/10.1007/978-981-16-2278-6_10

reported optimal settings of machining condition which yields superior performance in machining of aluminum matrix composites. It was reported a boost in cutting speed (3.234%), kerf width (2.7415%), and surface roughness (7.053%). Rajmohan and Senthil [10] have investigated the effect of pulse on time, travel speed, and current on material removal rate, surface finish, and cutting width in the machining of grade 2205 nitrogen-enhanced duplex stainless. Manoj et al. [11, 12] have highlighted the influence of cutting speed during taper machining on various parameters like surface roughness, slant angle, etc., and its effect corner radius during slant profiling. Sharma et al. [13] investigated the effect of WEDM machining parameters on the turbine disk profile.

Oblique and tapered shaped components can also be machined using WEDM, as it has many precision applications [14, 15]. Kinoshita et al. [16] highlighted the drawbacks of the tapering process in WEDM. They also examined the behavior of the wire during taper machining and studied the deviation of machined angle from the programmed angle. Yan et al. [17] designed two types of mechanism to avoid wear, angular error in case of tapering in WEDM. A new guide system was suggested due to the flaws in the designed mechanism, to reduce roundness error without damaging the surface of the workpiece. Sanchez et al. [18] have formulated a FEM model considering the nonlinear phenomenon of the wire to predict and reduce the angular error during tapering in WEDM. With the aid of prediction models, deviations of accuracies were reduced below $4'$ in tapered parts machined by WEDM. Plaza et al. [19] have modeled for prediction of angular error for reducing the experimentation in WEDM. The reduction of angular error below $3'45''$ was reported using the prediction model. Ranjit et al. [20] have reported the variation of cutting speeds and its influence on recast layer thickness during taper machining in WEDM.

From the literature, it was highlighted cutting speed (profiling speed) was an important parameter as major researchers have considered for investigation. Analyzing the profiling speed during taper machining in WEDM was very essential. In the present studies, a fixture was used to avoid disadvantages of conventional tapering by WEDM like guide wear, wire bending, insufficient flushing and inaccuracies in taper, etc. Almost all literature concentrates on machining parameters affecting profiling speed; but in this paper, an attempt is made to show that the geometric profile shapes also affects the profiling speed. The effects and variation of profiling speed were investigated while machining different profiles like triangular, square, and circular. The input parameters were wire guide distance (WGD), corner dwell time (CDT), wire offset (WO), and cutting speed override (CSO) for all the profiles. Different profiles of 1, 3, and 5 mm were machined at two different slant angles, namely 0° and 30°. All the profiles were machined using $L_{16}$ Taguchi orthogonal array at two different slant angles. It was found that CSO has a major influence on profiling speed and different profiles geometry also affected profiling speed.

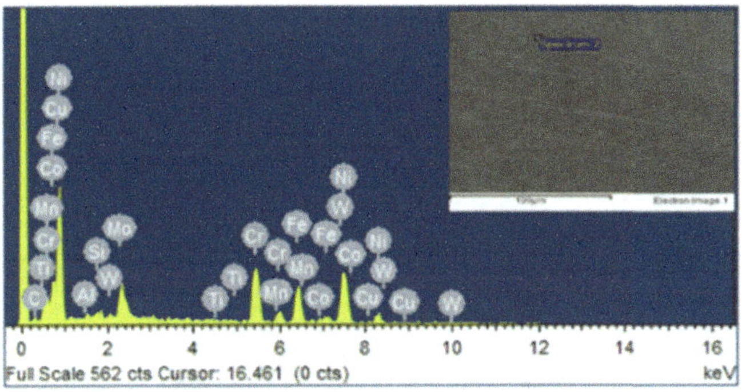


| Element | Weight% |
|---|---|
| C | 7.75 |
| Al | 0.41 |
| Si | 0.11 |
| Ti | 0.13 |
| Cr | 18.33 |
| Mn | 0.75 |
| Fe | 17.34 |
| Co | 0.48 |
| Ni | 44.53 |
| Cu | 1.41 |
| Mo | 7.83 |
| W | 0.95 |
| Totals | 100.00 |

**Fig. 1** Energy dispersive X-ray analysis of Hastelloy X

# 2 Material and Experimental Procedure

## *2.1 Machining Parameters*

Hastelloy X is a nickel-based superalloy having high mechanical properties, oxidation resistance, fabricability, and high-temperature strength. It has many applications in various industries like defence, aerospace, petrochemical, and dies making. The composition of the as-received alloy is shown as in Fig. 1 which was examined by energy dispersive X-ray analysis (EDX). It was solution heat-treated at 2150 °F (1177 °C) and rapidly cooled holding time of 1 h per inch of the section.

## *2.2 Experimental Particulars*

The dimension of the profile that was programmed was as shown in Fig. 1a, b using ELPULS software, and the necessary G&M codes were generated based on machining conditions. The generated codes were exported to Electronica 'ELPULS 15 CNC WEDM' which was used to machining the material. The zinc coated copper wire electrode and deionized water as dielectric fluid were employed through the experimental runs. In WEDMs, the sparks are generated between the wire and workpiece due to voltage difference which melts the material that is required to the machine. The dielectric fluid cools the workpiece and flushes the debris outside the machining area. The machining was performed on the workpiece at different slant angles as shown in Fig. 2a. The slant machining was achieved by the unique fixture that is made of aluminum H9. The workpiece was fixed to the angle plate which would be rotated for the required slant angle. This angle plate would be locked with the help of slots on the side plate to a required angle. The whole fixture is fixed on the table of WEDM as shown in Fig. 2b.

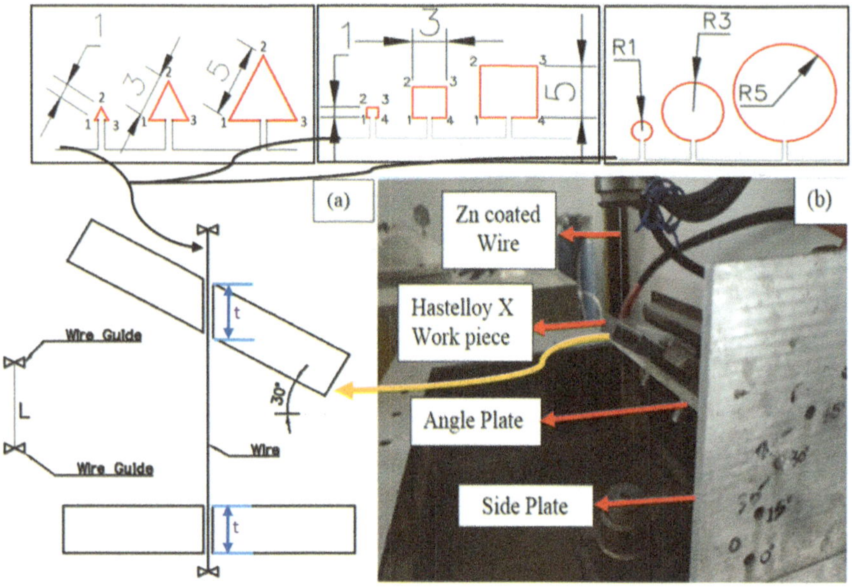


**Fig. 2** **a** Workpiece at different slant angle by slant fixture, **b** slant fixture

## 2.3 Machining Parameters

The initial trial experiments were conducted the machining parameters and constant cutting parameters were chosen as shown in Table 1 [11, 12]. Based on the trial runs, the levels were fixed for input parameters like WGD, CDT, WO, and CSO as shown

**Table 1** WEDM parameters

| EDM parameters | Settings | | | |
|---|---|---|---|---|
| *Constant parameters* | | | | |
| Pulse off time (μs) | 44 | | | |
| Servo feed (mm/min) | 20 | | | |
| Wire feed (m/min) | 6 | | | |
| Pulse on time (μs) | 115 | | | |
| Servo voltage (V) | 40 | | | |
| *Machining parameters* | | | | |
| Wire guide distance (mm) (WGD) | 40 | 50 | 60 | 70 |
| | 100 | 110 | 120 | 130 |
| Cutting speed override (%) (CSO) | 31 | 54 | 77 | 100 |
| Wire offset (μm) (WO) | 0 | 40 | 80 | 120 |
| Corner dwell time (s) (CDT) | 0 | 33 | 66 | 99 |

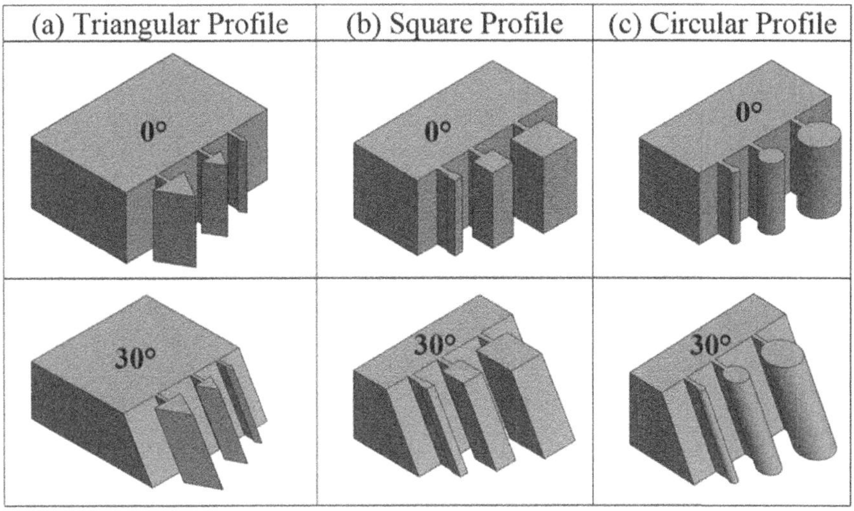


**Fig. 3** Machined **a** triangular, **b** square, and **c** circular taper profiles

in Table 1. Taguchi's L16 orthogonal array was employed in the present investigation based on different parametric level set for slant angles at 0°, and 30°. Figure 3a–c shows the different profiles machined of 10 mm thickness at different slant angles.

## 3 Results and Discussion

### *3.1 Profiling Speed of Different Profiles at 0° and 30° Angle*

The profiling speed was recorded during profiling provided by WEDM during profiling all the profiles, and the average of the values is taken as profiling speed. Table 2 shows the profiling speed calculated for triangular, square, and circular forms at 0° and 30° slant angles, respectively. It can be observed that the trials 12 and 28 yielded the least profiling speed and trials 15 and 31 produced the maximum profiling speed for all the slant angles. It can be observed that the square profile produced higher profiling speed than the circular and triangular profiles.

### *3.2 ANOVA and Main Effects Plots*

Table 3 shows ANOVA for different profiles at 0° and 30° slant angles, respectively. It can be clearly noticed that the CSO parameter was the most significant factor influencing the profiling speed. The ANOVA also showed that CSO contributed the

**Table 2** Profiling speed of different profiles at 0° and 30° angle

| Trial No | WGD (mm) | CDT (s) | WO (μm) | CSO (%) | Profiling speed | | |
|---|---|---|---|---|---|---|---|
| | | | | | Triangle (mm/min) | Square (mm/min) | Circular (mm/min) |
| 0° | | | | | | | |
| 1 | 40 | 0 | 0 | 31 | 0.749 | 0.769 | 0.756 |
| 2 | 40 | 33 | 40 | 54 | 0.952 | 1.126 | 0.990 |
| 3 | 40 | 66 | 80 | 77 | 1.256 | 1.427 | 1.392 |
| 4 | 40 | 99 | 120 | 100 | 1.425 | 1.616 | 1.537 |
| 5 | 50 | 0 | 40 | 77 | 1.266 | 1.501 | 1.337 |
| 6 | 50 | 33 | 0 | 100 | 1.409 | 1.599 | 1.510 |
| 7 | 50 | 66 | 120 | 31 | 0.659 | 0.898 | 0.748 |
| 8 | 50 | 99 | 80 | 54 | 0.801 | 1.000 | 0.881 |
| 9 | 60 | 0 | 80 | 100 | 1.403 | 1.558 | 1.501 |
| 10 | 60 | 33 | 120 | 77 | 1.371 | 1.523 | 1.398 |
| 11 | 60 | 66 | 0 | 54 | 0.812 | 0.890 | 0.879 |
| 12 | **60** | **99** | **40** | **31** | **0.631** | **0.740** | **0.718** |
| 13 | 70 | 0 | 120 | 54 | 0.750 | 0.908 | 0.898 |
| 14 | 70 | 33 | 80 | 31 | 0.602 | 0.750 | 0.729 |
| 15 | **70** | **66** | **40** | **100** | **1.508** | **1.719** | **1.677** |
| 16 | 70 | 99 | 0 | 77 | 1.326 | 1.495 | 1.393 |
| 30° | | | | | | | |
| 17 | 100 | 0 | 0 | 31 | 0.483 | 0.695 | 0.620 |
| 18 | 100 | 33 | 40 | 54 | 0.566 | 0.718 | 0.703 |
| 19 | 100 | 66 | 80 | 77 | 1.000 | 1.160 | 1.106 |
| 20 | 100 | 99 | 120 | 100 | 1.384 | 1.467 | 1.411 |
| 21 | 110 | 0 | 40 | 77 | 0.977 | 1.297 | 1.089 |
| 22 | 110 | 33 | 0 | 100 | 1.361 | 1.499 | 1.383 |
| 23 | 110 | 66 | 120 | 31 | 0.494 | 0.698 | 0.682 |
| 24 | 110 | 99 | 80 | 54 | 0.571 | 0.779 | 0.718 |
| 25 | 120 | 0 | 80 | 100 | 1.383 | 1.450 | 1.405 |
| 26 | 120 | 33 | 120 | 77 | 1.046 | 1.369 | 1.074 |
| 27 | 120 | 66 | 0 | 54 | 0.586 | 0.834 | 0.696 |
| 28 | **120** | **99** | **40** | **31** | **0.470** | **0.693** | **0.607** |
| 29 | 130 | 0 | 120 | 54 | 0.586 | 0.856 | 0.708 |
| 30 | 130 | 33 | 80 | 31 | 0.500 | 0.698 | 0.614 |
| 31 | **130** | **66** | **40** | **100** | **1.450** | **1.469** | **1.457** |
| 32 | 130 | 99 | 0 | 77 | 1.006 | 1.406 | 1.088 |

The lowest and highest profiling parameters are indicated in the bold format

**Table 3** ANOVA at for different profiles geometries at 0° and 30° angle

| Sl No. | Factor | DF | Adj SS | Adj MS | *F*-value | *P*-value | % contribution |
|---|---|---|---|---|---|---|---|
| 0° | | | | | | | |
| *Triangular profile* | | | | | | | |
| 1 | WGD | 3 | 0.009 | 0.003 | 0.34 | 0.798 | 0.521 |
| 2 | CDT | 3 | 0.004 | 0.001 | 0.15 | 0.921 | 0.233 |
| 3 | WO | 3 | 0.012 | 0.004 | 0.47 | 0.725 | 0.711 |
| 4 | **CSO** | **3** | **1.656** | **0.552** | **63.92** | **0.003** | **97.017** |
| 5 | Error | 3 | 0.026 | 0.009 | | | |
| 6 | Total | 15 | 1.706 | | | | |
| *Square profile* | | | | | | | |
| 1 | WGD | 3 | 0.012 | 0.004 | 0.40 | 0.761 | 0.582 |
| 2 | CDT | 3 | 0.010 | 0.003 | 0.34 | 0.802 | 0.486 |
| 3 | WO | 3 | 0.021 | 0.007 | 0.74 | 0.595 | 1.065 |
| 4 | **CSO** | **3** | **1.905** | **0.635** | **67.03** | **0.003** | **96.428** |
| 5 | Error | 3 | 0.028 | 0.010 | | | |
| 6 | Total | 15 | 1.976 | | | | |
| *Circular profile* | | | | | | | |
| 1 | WGD | 3 | 0.010 | 0.003 | 1.21 | 0.440 | 0.559 |
| 2 | CDT | 3 | 0.006 | 0.002 | 0.77 | 0.581 | 0.357 |
| 3 | WO | 3 | 0.007 | 0.002 | 0.83 | 0.559 | 0.383 |
| 4 | **CSO** | **3** | 1.778 | **0.593** | **212.72** | **0.001** | **98.240** |
| 5 | Error | 3 | 0.008 | 0.003 | | | |
| 6 | Total | 15 | 1.810 | | | | |
| 30° | | | | | | | |
| *Triangular profile* | | | | | | | |
| 1 | WGD | 3 | 0.003 | 0.0009 | 1.14 | 0.459 | 0.136 |
| 2 | CDT | 3 | 0.002 | 0.0006 | 0.69 | 0.617 | 0.082 |
| 3 | WO | 3 | 0.001 | 0.0003 | 0.30 | 0.823 | 0.035 |
| 4 | **CSO** | **3** | **2.106** | **0.702** | **838.11** | **0.0001** | **99.628** |
| 5 | Error | 3 | 0.003 | 0.0009 | | | |
| 6 | Total | 15 | 2.114 | | | | |
| *Square profile* | | | | | | | |
| 1 | WGD | 3 | 0.02099 | 0.007 | 13.38 | 0.030 | 1.181 |
| 2 | CDT | 3 | 0.00462 | 0.002 | 2.94 | 0.199 | 0.260 |
| 3 | WO | 3 | 0.02085 | 0.007 | 13.30 | 0.031 | 1.174 |
| 4 | **CSO** | **3** | **1.72868** | **0.576** | **1102.48** | **0.0001** | **97.297** |
| 5 | Error | 3 | 0.00157 | 0.0005 | | | |

(continued)

**Table 3** (continued)

| Sl No. | Factor | DF | Adj SS | Adj MS | *F*-value | *P*-value | % contribution |
|---|---|---|---|---|---|---|---|
| 6 | Total | 15 | 1.77671 | | | | |
| *Circular profile* | | | | | | | |
| 1 | WGD | 3 | 0.001 | 0.0004 | 1.14 | 0.458 | 0.081 |
| 2 | CDT | 3 | 0.004 | 0.0013 | 3.38 | 0.172 | 0.238 |
| 3 | WO | 3 | 0.001 | 0.0004 | 0.96 | 0.513 | 0.067 |
| 4 | **CSO** | **3** | **1.582** | **0.5274** | **1412.68** | **0.0001** | **99.543** |
| 5 | Error | 3 | 0.001 | 0.0004 | | | |
| 6 | Total | 15 | 1.590 | | | | |

CSO is the highest contributing and significant factor so it is indicated in bold format

highest irrespective of slant angle and profile geometry. Figures 4 and 5 show the main effects plots for triangular, square, and circular profiles at both angles. It can be concluded that CSO is the most influencing factor on profiling speed, and as the CSO percentage increases, the profiling speed also increases. In Figs. 4 and 5 at 54–77%, there is variation in profiling speed. This increase was observed due to the flexible nature of wire which vibrates and leads to variation in the profiling speed [21, 22], whereas other parameters like WGD, CDT, and WO has very minimal effect

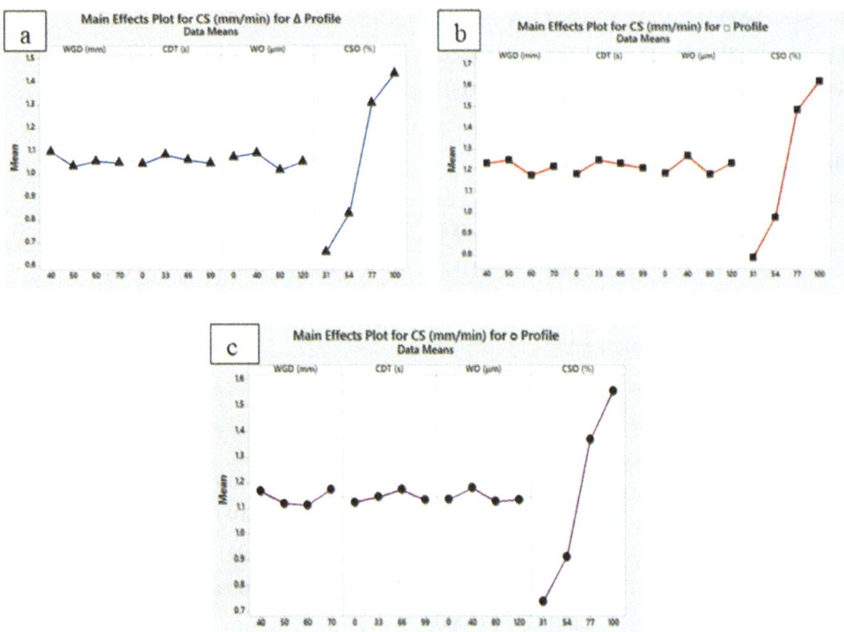


**Fig. 4** Main effects plots for **a** triangular, **b** square, and **c** circular profiles at 0°

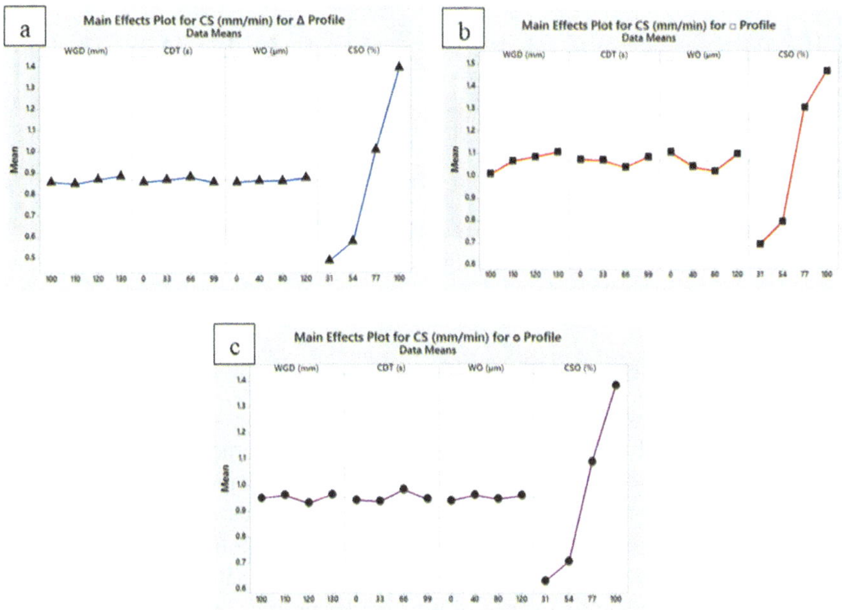


**Fig. 5** Main effects plots for **a** triangular, **b** square, and **c** circular profiles at 30°

on profiling speed irrespective of angle and profile geometry. There are very small variations observed which is due to errors during machining. It can also be noticed that the highest profiling speed parameter has the highest CSO percentage (100%), and lowest profiling speed parameter has the lowest CSO percentage (31%). The profiling speed also decreased as the angles increased from 0° to 30° because of the increase in cutting thickness.

### *3.3 Variation of Profiling Speed for Different Profiles Geometry*

Figure 6 shows the variation of profiling speed at 0° and 30° slant angle, respectively. It can be observed that the triangular profile has the minimum profiling speed, and the square profile had the maximum profiling speed. During machining the profiles, the instantaneous profiling speed increases or decreases gradually at the corners of the profile, and then, it comes back to the original profiling speed at straight edge cuts due to the CSO parameter. The CSO parameter detects the corners (acute-angled corners or right-angled corners) in the profile geometry and changes conditions during machining. It automatically tunes the speed to increases or decreases for obtaining the sharp corners in case of square and triangular profiles without changing the machining parameters [23]. Acute-angled corners (triangular shape) need more

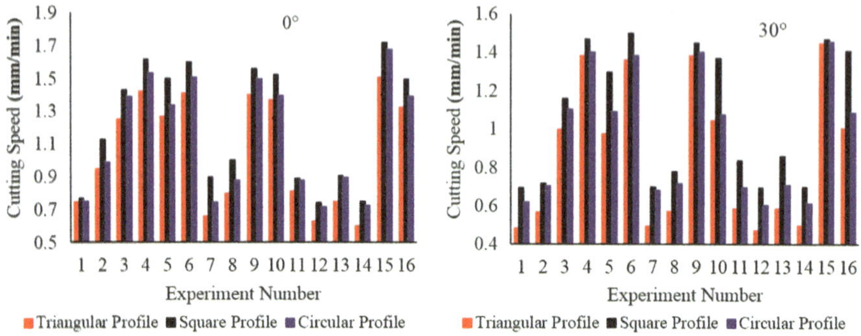


**Fig. 6** Variation of profiling speed for different profiles

time and lesser speed for its machining in order to avoid corner errors [23, 24]. So, the square profiles have the highest profiling speed due to geometry, as it has a higher number of corners compared to triangular profiles. In the case of circular profiles, the wire has to traverse by changing the direction at every point. This movement of the wire in a curved way changes the direction very instant that decreases the cutting speed. This makes the cutting speed of the circular profiles lower compared to the square profile. But comparing the triangular profile, the circular profile has a higher cutting speed due to the geometry of the profile and CSO parameter. Similar trends are found all the profiles which were machined at 30° slant angle.

## 4 Conclusions

Form the above investigation, the CSO parameter was analyzed for different profile geometries at 0° and 30° slant angles. The following conclusions are drawn.

1. The CSO parameter is the most significant and highest contributing factor (96.428–99.628%) that influences and controls the profiling speed for all the profile at both 0° and 30° slant angles.

2.All the profiles indicated that profiling speed decreased as the slant angle increased from 0° to 30°. The greatest decrease of 40.55%, 36.26%, and 28.98% was observed in triangular, square, and circular profiles, respectively.

3.For the same parameters, there was an increase of 1.29–31.54% observed in profiling speed due to its geometry. The triangular profile was found to have least profiling speed compared to all profiles.

## References


1. Thakur A, Gangopadhyay S (2016) State of the art in surface integrity in machining of nickel based super alloys. Int J Mach Tools Manuf 100:25–54
2. Pramanik A (2014) Problems and solutions in machining of titanium alloys. Int J Adv Manuf Technol 70:919–928
3. Okunkova A, Peretyagin P, Seleznyov A, Fedorov SV, Kozochkin M (2016) Characterization of material's defects after electrical discharge machining and research into their technological parameters using vibroacoutstic diagnostics. Adv Mater Lett 7(7):542–548
4. Pramanik A (2016) Electrical discharge machining of MMCs reinforced with very small particles. Mater Manuf Process 31:397–404
5. Manoj IV, Narendranath S (2020) Variation and artificial neural network prediction of profile areas during slant type taper profiling of triangle at different machining parameters on Hastelloy X by wire electric discharge machining. Proc Inst Mech Eng Part E J Process Mech Eng 234(6):673–683
6. Pramanik A, Islam MN, Boswell B, Basak AK, Dong Y, Littlefair G (2018) Accuracy and finish during wire electric discharge machining (EDM) of metal matrix composites (MMCs) for different reinforcement size and machining conditions. Proc Inst Mech Eng Part B J Eng Manuf 232(6):1068–1078
7. Mouralova K, Benes L, Bednar J, Zahradnicek R, Prokes T, Matousek R, Hrabec P, Fiserova Z, Otoupalik J (2019) Using a DoE for a comprehensive analysis of the surface quality and cutting speed in WED-machined hadfield steel. J Mech Sci Technol 33(5):2371–2386
8. He W, He S, Du J, Ming W, Ma J, Cao Y, Li X (2019) Fiber orientations effect on process performance for wire cut electrical discharge machining (WEDM) of 2D C/SiC composite. Int J Adv Manuf Technol 102:507–518
9. Dey A, Pandey KM (2018) Selection of optimal processing condition during WEDM of compocasted AA6061/cenosphere AMCs based on grey-based hybrid approach. Mater Manuf Processes 33(14):1549–1558
10. Rajmohan K, Senthil Kumar A (2017) Experimental investigation and prediction of optimum process parameters of micro-wire-cut EDM of 2205 DSS. Int J Adv Manuf Technol 93:187–201
11. Manoj IV, Joy R, Narendranath S (2019) Investigation on the effect of variation in cutting speeds and angle of cut during slant type taper cutting in WEDM of Hastelloy X. Arab J Sci Eng 45(2):641–651
12. Manoj IV, Joy R, Narendranath S, Nedelcu D (2019) Investigation of machining parameters on corner accuracies for slant type taper triangle shaped profiles using WEDM on Hastelloy X. IOP Conf Ser Mater Sci Eng 591(1):1–11
13. Sharma P, Chakradhar D, Narendranath S (2015) Evaluation of WEDM performance characteristics of Inconel 706 for turbine disk application. Mater Des 88:558–566
14. Selvakumar G, Jiju KB, Sarkar S, Mitra S (2016) Enhancing die corner accuracy through trim cut in WEDM. Int J Adv Manuf Technol 83:791–803
15. Mussada EK, Hua CC, Rao AKP (2018) Surface hardenability studies of the die steel machined by WEDM. Mater Manuf Processes 33(16):1745–1750
16. Kinoshita N, Fukui M, Fujii T (1987) Study on wire-EDM: accuracy in Taper-Cut. Annal CIRP 36:119–122
17. Yan H, Liu Z, Li L, Li C, He X (2017) Large taper mechanism of HS-WEDM. Int J Adv Manuf Technol 90:2969–2977
18. Sanchez JA, Plaza S, Ortega N, Marcos M, Albizuri J (2008) Experimental and numerical study of angular error in WEDM taper–cutting. Int J Mach Tools Manuf 48:1420–1428
19. Plaza S, Ortega N, Sanchez JA, Pombo I, Mendikute A (2009) Original models for the prediction of angular error in wire-EDM taper-cutting. Int J Adv Manuf Technol 44:529–538
20. Manoj IV, Joy R, Narendranath S (2019) Investigation of cutting speed, recast layer and micro-hardness in angular machining using slant type taper fixture by WEDM of Hastelloy X. Mater Today Proc 27:1943–1946

21. Habib S, Okada A (2016) Experimental investigation on wire vibration during fine wire electrical discharge machining process. Int J Adv Manuf Technol 84:2265–2276
22. Habib S (2017) Optimization of machining parameters and wire vibration in wire electrical discharge machining process. Mech Adv Mater Mod Process 3(3):1–9
23. Electronica India Ltd. (2011) Operating manual for ELPLUS 15 Ecocut, 2011 Retrieved by email from company. Access date 25–06–2011
24. Bisaria H, Shandilya P (2019) Processing of curved profiles on Ni-rich nickel–titanium shape memory alloy by WEDM. Mater Manuf Processes 34(12):1333–1341

# Evaluation of Machining Properties of Short Bamboo Fiber-Based Green Composites Using CNC Drilling Process

**Shubham B. Patil, Jagadish, Shailesh Vaidya, and Satish Kumar Adapa**

**Abstract** Today's, use of natural fiber-based polymer composites is also known as green composites in various industrial applications due to their superior properties. Machining of these composite materials is one of the important operations while manufacturing different engineered components. The machining of NFRP composites bit difficult and differs from other composites. The aim of the paper is to machining study of short bamboo fiber-based green composites (SBFGC) in dry conditions using CNC drilling process. Initially, composite with 2.5% bamboo fiber +97.5% polylactic acid (PLA) is fabricated using compression molding method. Then, experimentations are performed by varying the input parameters [drill bit speed (DBS) and cutting speed (CS)] as per the Taguchi (L9) OA and corresponding output parameter like thrust force and material removal rate (MRR) are determined. Further, parametric effect analysis and ANOVA are performed for determining the influencing parameters and statistical significance of the data, respectively. The result shows that the parameters DBS and CS have contributed more effect for thrust force (TF) while input parameter CS for MRR during the machining of SBFGC. At last, confirmatory test results show comparable and acceptable with the experimental results.

**Keywords** SBFGC · CNC drilling process · Parametric analysis · ANOVA · Taguchi method

S. B. Patil · Jagadish (✉) · S. Vaidya
Department of Mechanical Engineering, National Institute of Technology Raipur, Chhattisgarh 492010, India
e-mail: jagadish.mech@nitrr.ac.in

S. Vaidya
e-mail: svaidya.ws@nitrr.ac.in

S. K. Adapa
Department of Mechanical Engineering, Aditya Institute of Technology and Management College, Tekkali, Andhra Pradesh 532201, India



P. Srinivasa Pai and V. Krishnaraj (eds.), *Sustainable Machining Strategies for Better Performance*, Lecture Notes in Mechanical Engineering,
https://doi.org/10.1007/978-981-16-2278-6_11

# 1 Introduction

Natural fibers reinforced composites (NFRC) have great potential to substitute synthetic material. These are lignocelluloses fibers that are extracted from different plants like bamboo, flax, jute, banana, kenaf, and hemp. Bamboo is a glass-like plant of the *Bambusoideae* family. Bamboo fiber is also considered the natural fiber owing to its lightweight and strength ratio. Bamboo fiber-based polymer composite is used in automobile, aerospace, building construction, and packaging industries.

Instead, bamboo fibers are eco-friendly, less in density and economically cheap and have larger industrial application compare to the other fibers [hemp, coir, flax, coconut, sisal and banana, bagasse's, rice husk, etc.]. Meanwhile, some extent of work has been done on the bamboo fiber-based composites with their physical and mechanical properties. Anwar et al. determine the physical and mechanical properties of phenolic-treated bamboo strips. The result shows that, the phenolic-treated strips have greatly improved their properties [1]. Kamruzzaman et al. study the effect of height and age on mechanical property of bamboo. The modulus of elasticity (MOE) and the rupture modulus not affected by height and weight [2]. Verma and Charier measure the strength of the bamboo epoxy laminate composite as tensile, compressive, and flexural. The results shows that increased tensile and compression properties decrease in lamina angle. Lamina configuration has majorly affected the mechanical properties [3].

On the other hand, a SBFGC composite undergoes various secondary operations before going to the actual use. Secondary operations include like drilling, trimming, cutting, and milling in order to get the final product. ElayaPerumal and Venkateshwaran studied delamination analysis at composite plate's entry and exit depending on the parameters of the drilling process. They performed drilling on banana by altering speed and feed of machine. The results showed that, the speed is more influencing factor than feed rate on delamination [4]. Babu et al. explore the effects of hemp fiber composite machining. In order to obtain minimal conditions for delamination, the Taguchi technique was used to plan the experiments and ANOVA analysis. To reduce the various kinds of damage, cutting speed and feed rate need to be selected very carefully [5]. Naveen et al. [6] examine the effect of speed and feed, on the damage factor of hemp and glass composite fibers with specific fraction of fiber volume (i.e., 10, 20, and 30%). It has been identified that higher feed levels cause the damage around the hole opening. The best results for the drilling of FRP composite laminates are high cutting speeds (40 and 60 rpm) and lower feed rates (0.1, 0.2 mm/rev). Wang et al. analyze the results of the conventional hemp fiber composite laminate drilling (CD) and hybrid ultrasound drilling (UAD).The results show that UAD is effective for a variety of drilling conditions in comparison with CD [7]. Jayabala et al. experimentally investigate machinability characteristic of coir reinforced polymer composites (CRPC). For CRPC, characteristic analysis by Taguchi regression models has been established. The optimum value of thrust force, torque, and tool wear in drill analysis was determined by the 6 mm drill diameter, the 600 rpm spindle speed, and the feed rate by 0–3 mm/rev. [8]. Yallew et al. experimentally investigate effect of thrust force and torque on jute reinforced PP composite. The test values show a close relation between the geometry of the device and the delamination factor [9]. Kumar and Naidu study the machinability characteristics of coconut and banana fiber-based composite. The feed rate and depth of cut have direct relationship while speed has inverse relationship on milling operation. Depth of cut 0.5 mm, feed rate 60 mm/min, and speed 800 rpm are optimum values [10].

Moreover, processing/machining of SBFBGC using CNC drilling process has been not reported so far in the literature. Here, first, short bamboo fibers are extracted from the culm of bamboo using a novel mechanical process. Further, the SBFBGC is fabricated with 2.5% composition using compression molding process. Thereafter, an experimental investigation is carried out based on Taguchi (L9-OA) to analyze the machining attributes [such as material removal rate (MRR), and thrust force (TF)] to the variation in process parameters like DBS and CS. In addition, parametric analysis has to be done to show the statistical significance of the CNC drilling process. At last, confirmatory tests are performed to verify the experimental results.

## 2 Materials and Method

### *2.1 Extraction and Manufacturing of SBF-Based Composites*

The short bamboo fiber (SBF) was extracted using mechanical processes. The detail steps of extraction and manufacturing of SBF-based polymer composites are depicted in Fig. 1. First, bamboo stakes are collected from local market in Raipur, Chhattisgarh, and then outer and inner layers of the bamboo stakes are peeled off [11]. Bamboo culm cut into small pieces using hand saw machine. Second, bamboo pieces are attached on lathe machine and perform facing operation. Due to facing operation, various size fibers cut from the bamboo pieces and collected in the beaker. Third step, different size particles are segregated using sieve plate also known as sieving operation. Fourth, different chemical treatments like NaOH solution and acetone solution are done to remove the impurities presence in the SBF's [12]. Fifth, 2.5% of SBF and 97.5% of polylactic acid (PLA) is stirred properly using mechanical stirrer and poured into the glass mold of size 100 mm × 50 mm × 5 mm. The glass mold is kept for 24 h at room temperature for curing of mixer. At last, prepared samples of SBF-based polymer composites of size 100 mm × 50 mm × 5 mm are taken for machining.

### *2.2 Machining Procedure*

The machining is performed in CNC drilling center manufactured by MTAB Pvt Ltd as shown in Fig. 2. The two major input parameters, namely drill bit speed (DBS) and

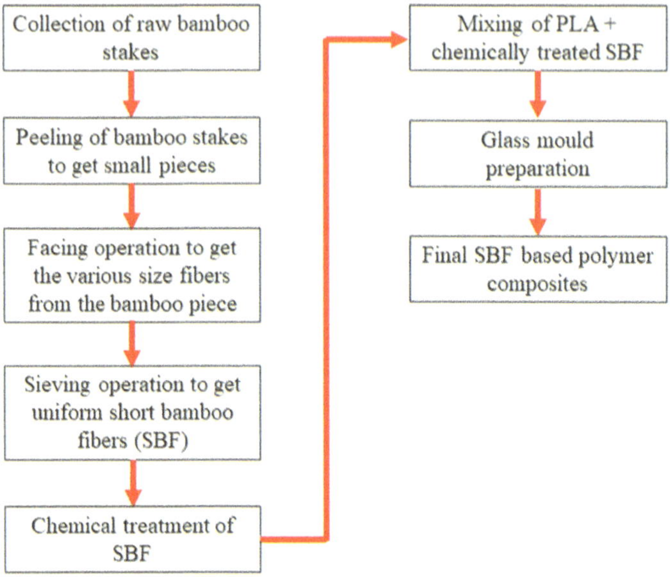


**Fig. 1** Extraction and manufacturing steps of SBF-based polymer composite

**Fig. 2** CNC drilling machine

cutting speed (CS), are used and corresponding output parameter like thrust force and MRR are measured. The details of input parameter and their levels are given in Table 1.

**Table 1** Input parameter and their levels for CNC machining center

| Input parameters | Symbol | Units | Level 1 | Level 2 | Level 3 |
|---|---|---|---|---|---|
| Drill bit speed | DBS | rpm | 500 | 800 | 1200 |
| Cutting speed | CS | mm/min | 60 | 120 | 180 |
| *Constant parameters* | | | | | |
| Drill bit diameter: 12 mm | | | | | |
| Tool material: HSS | | | | | |
| Coolant: No coolant | | | | | |

During the experimentation, two input parameters DBS and CS are varied according to the Taguchi (L9) orthogonal array and a hole of Ø 12 mm are drilled. Each experiment is performed 3 times, and their average of thrust force (TF) and material removal rate (MRR) are taken for the analysis [13, 14]. The parameter and thrust force (TF) values are obtained for each of the experiment using a dynamometer which was made using a 40 kg load cell, HX711 module and Uno-Arduino. The parameter MRR is determined using following expression.

$$\text{MRR} = \frac{I_m - F_m}{\rho \times \text{CT}} \tag{1}$$

where $I_m$ and $F_m$ are the initial and final weights of SBFGC composite; $\rho$ is density; CT is cutting time; MRR is material removal rate. $I_m$ and $F_m$ are measured by using precision weighing balance machine with an accuracy of ±0.1. The experimental results of CNC drilling process are given in Table 2.

**Table 2** Experimental results of 2.5% SBFGC composites of CNC drilling

| S. No. | Drill bit speed (rpm) | Cutting speed (mm/min) | Thrust force (N) | MRR ($mm^3$/s) |
|---|---|---|---|---|
| 1 | 500 | 60 | 122.41 | 100.14 |
| 2 | 500 | 120 | 129.23 | 151.21 |
| 3 | 500 | 180 | 131.74 | 297.20 |
| 4 | 800 | 60 | 102.25 | 99.56 |
| 5 | 800 | 120 | 109.32 | 153.24 |
| 6 | 800 | 180 | 115.45 | 271.36 |
| 7 | 1200 | 60 | 65.32 | 100.51 |
| 8 | 1200 | 120 | 70.65 | 160.32 |
| 9 | 1200 | 180 | 79.24 | 300.41 |

# 3 Results and Discussions

## 3.1 Parametric Analysis

The plot (Fig. 3) shows the effect of each input process (DBS, CS parameters on TF of SBFGC. It is observed that, TF is decreased from 127 to 71 N when the term DBS from increased from 500 to 1200 rpm. This is due to the fact that, as the drill bit speed increased from 500 to 1200 rpm, the impact of tool force on the SBFGC is less, i.e., required less force to remove the materials because of more speed of drill bit; hence, results show that decrement of TF w.r.t to the DBS. On the other hand, opposite behavior of TF is shown in case of CS when it increased from 60 to 180 mm/min, the TF is increased from 96 to 108 N. This is due to the fact that, as the CS increased the feed rate of the tool increased which makes the drill bit to enforce into the SBFGC results in higher TF [15]. Based on the parametric analysis, the optimal setting obtained for lower TF as DBS (1200 rpm, level 3) and CS (60 mm/min, level 1).

The main effect plot of MRR of SBFGC is shown in Fig. 4. From the main effect plot, it is observed that, MRR is first decreased from 182 to 174 $mm^3/s$ when the DBS increased from 500 to 800 rpm. Then there is sudden increment of MRR from 174 to 187 $mm^3/s$ when the DBS increased from 800 to 1200 rpm. This may the reason due to the sudden increment of drill bit speed results in reduction of TF which ultimately reduction in MRR. However, further increment of speed gave higher MRR as per the physics of the machining process. With respect to the CS, the rate MRR is significantly increased from 100 to 289 $mm^3/s$ when the term CS is increased from 60 to 180 mm/min. This is due to the fact that, as the CS increased means, penetration forces of the drill bit increase which results in cutting of more

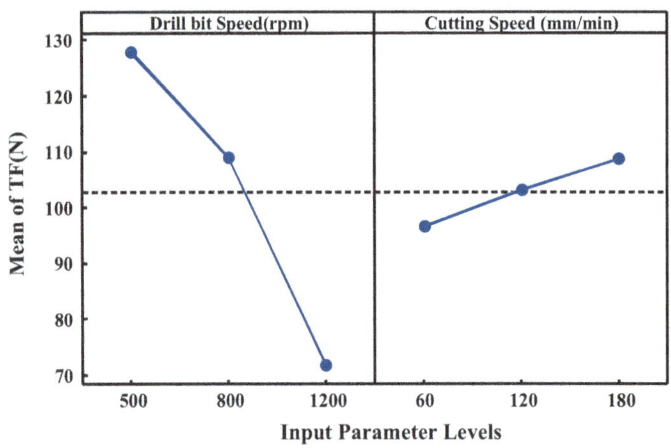


**Fig. 3** Main effect plot of DBS and CS on TF of SBFGC

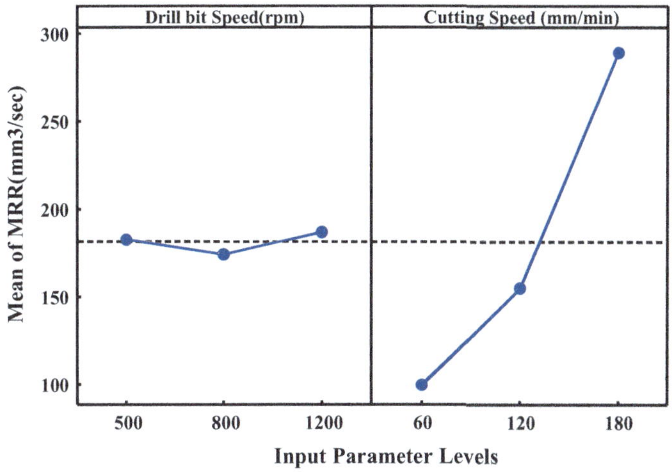


**Fig. 4** Main effect plot of DBS and CS on MRR of SBFGC

materials from the workpiece [14, 15]. Hence, in the case of SBFGC machining, the MRR found greater when the feed of the tool and speed of the tool when increases. Based on the parametric analysis, the optimal setting obtained for lower TF as DBS (1200 rpm, level 3) and CS (180 mm/min, level 3).

## *3.2 ANOVA Analysis*

An analysis of variance of data is conducted for TF and MRR with the goal of evaluating the effects and statistical significance of DBS and CS. The ANOVA analysis is performed at 95% confidence level. The result of ANOVA for TF and MRR is given in Tables 3 and 4. Higher $F$ value and lower $p$ values of each parameter show the statistical significance of that parameter. It is observed that, the parameters DBS and CS are statistical significance for TF while only parameter CS is found statistical

**Table 3** ANOVA of TF SBFGC

| Source | DF | Adj SS | Adj MS | *F* value | *P* value |
|---|---|---|---|---|---|
| Regression | 2 | 5051.28 | 2525.64 | 230.99 | 0 |
| DBS | 1 | 4829.85 | 4829.85 | 441.73 | 0 |
| CS | 1 | 221.43 | 221.43 | 20.25 | 0.004 |
| Error | 6 | 65.6 | 10.93 | | |
| Total | 8 | 5116.88 | | | |

*R*-sq: 98.72%; *R*-sq(adj): 98.29%; *R*-sq(pred): 97.20%

**Table 4** ANOVA of MRR of SBFGC

| Source | DF | Adj SS | Adj MS | *F* value | *P* value |
|---|---|---|---|---|---|
| Regression | 2 | 53,955 | 26,977.5 | 43.71 | 0 |
| DBS | 1 | 40.4 | 40.4 | 0.07 | 0.807 |
| CS | 1 | 53,914.7 | 53,914.7 | 87.35 | 0 |
| Error | 6 | 3703.5 | 617.3 | | |
| Total | 8 | 57,658.6 | | | |

*R*-sq: 98.58%; *R*-sq(adj): 99.44%; *R*-sq(pred): 98.52%

significance for MRR due to lesser *P* value and higher *F* values. Hence, the parameters DBS and CS have more significant effect on TF and parameter CS on MRR. This justifies the results obtained from the parametric analysis. The value of *R*-sq and *R*-sq (adj) for both the output parameter, i.e., TF and MRR for CNC machining of SBFGC, is found to higher as (*R*-sq: 98.72%; *R*-sq (adj): 98.29% for TF) and (*R*-sq: 98.58%; *R*-sq (adj): 99.44%; for MRR). A higher value of *R*-sq and *R*-sq (adj) shows the better agreement of experimental values with respect to the normality of the data [15].

## 3.3 Confirmatory Experiments

Additionally, confirmatory tests are performed to verify the results obtained from the parametric analysis. The optimal setting for TF as DBS (1200 rpm, level 3) and CS (60 mm/min, level 1) and for MRR as DBS (1200 rpm, level 3) and CS (180 mm/min, level 3) is considered for the confirmatory experiments. The results of the confirmatory experiments for TF and MRR are given in Table 5. The result shows that, comparable and acceptable with the experimental results for the selected optimal setting.

**Table 5** Confirmatory tests for CNC drilling process of SBFGC

| Input conditions | Output parameters | Experimental results | Confirmatory tests results |
|---|---|---|---|
| *TF*:<br>DBS (1200 rpm, level 3) and CS (60 mm/min, level 1) | TF (N) | 71.73 | 72.43 |
| *MRR*:<br>DBS (1200 rpm, level 3) and CS (180 mm/min, level 3) | MRR ($mm^3/s$) | 289.65 | 287.34 |

## 4 Conclusions

Evaluation of machining properties, namely TF and MRR, of SBFGC in CNC drilling process is presented. Here, input parameter, namely DBS and CS, and output parameter TF and MRR are considered. Experiments are performed in CNC drilling setup based on Taguchi (L9) orthogonal array. The result shows that the parameter DBS and CS has contributed more effect for TF while input parameter CS for MRR during the machining of SBFGC. The optimal setting obtained for TF as DBS (1200 rpm, level 3) and CS (60 mm/min, level 1) and for MRR as DBS (1200 rpm, level 3) and CS (180 mm/min, level 3) to get lower TF and higher MRR in CNC drilling process. Additionally, the ANOVA for TF and MRR is carried out to determine the statistical significance of the input parameter of DBS and CS for CNC drilling process. At last, confirmatory test results show comparable and acceptable with the experimental results. Hence, it is concluded that, CNC drilling process is feasible and capable for machining of SBFGC. The optimal setting obtained from the parametric analysis can be used as the optimal setting for the CNC drilling process during the machining of SBFGC.

## References

1. Hojo T, Xu Z, Yang Y, Hamada H (2014) Tensile properties of bamboo, jute and kenaf mat-reinforced composite. Energy Procedia 56:72–79
2. Kamruzzaman M, Saha SK, Bose AK, Islam MN (2008) Effects of age and height on physical and mechanical properties of bamboo. J Trop For Sci, 211–217
3. Verma CS, Chariar VM (2012) Development of layered laminate bamboo composite and their mechanical properties. Compos B Eng 43(3):1063–1069
4. Venkateshwaran N, ElayaPerumal A (2013) Hole quality evaluation of natural fiber composite using image analysis technique. J Reinf Plast Compos 32(16):1188–1197
5. Babu GD, Babu KS, Gowd BUM (2012) Effects of drilling parameters on delamination of hemp fiber reinforced composites. Int J Mech Eng Res Dev 2:1–8
6. Naveen PNE, Yasaswi M, Prasad RV (2012) Experimental investigation of drilling parameters on composite materials. J Mech Civ Eng 2:30–37
7. Wang D, Onawumi PY, Ismail SO, Dhakal HN, Popov I, Silberschmidt VV, Roy A (2019) Machinability of natural-fibre-reinforced polymer composites: conventional vs. ultrasonically-assisted machining. Compos Part A: Appl Sci Manuf 119:188–195
8. Jayabal S, Natarajan U (2011) Drilling analysis of coir-fibre-reinforced polyester composites. Bull Mater Sci 34(7):1563–1567
9. Yallew TB, Kumar P, Singh I (2016) A study about hole making in woven jute fabric-reinforced polymer composites. Proc Inst Mech Eng, Part L: J Mater Des Appl 230(4):888–898
10. Kumar PK, Naidu KM (2019) Fabrication and machinability behavior of banana and coconut sheath fiber based composite. Int J Innovative Res Technol 6(2):206–216
11. Singh AS, Halder S, Wang J, Jagadish (2017) Extraction of bamboo micron fibers by optimized mechano-chemical process using a central composite design and their surface modification. Mater Chem Phys 199:23–33
12. Jagadish, Maran R, Amitava R (2018) Investigation on mechanical properties of pineapple leaf based short fiber reinforced polymer composite from selected Indian (north eastern part) cultivars. J Thermoplast Compos Mater 33(3):324–342

13. Binoy KB, Tapas D, Jagadish, Promod KP (2020) Machinability assessment of titanium grade 2 alloys using deionized water in EDM. Mater Today: Proc 26(2):2221–2225
14. Tapas D, Binoy KB, Jagadish, Promod KP (2020) Machinability study of 430 stainless steel using tap water in EDM. Mater Today: Proc 26(2):2179–2183
15. Sandeep K, Jagadish, Andeep KS, Narendra K (2020) Multi-objective optimization of CNC drilling parameters on HcHcr steel using Taguchi's based utility concept & GRA-PCA methods. Proceedings of International Conference on Advances in Mechanical Engineering (ICAME2020). Int J Adv Eng Sci Technol Res (IJAESTR), 55–63

# Influence of Burnishing Process on Tensile Strength of Al7075-T6 Alloy

**Pavana Kumara and Udaya Prasanna Handadi**

**Abstract** Burnishing is a finishing process that works on cold working principles and is performed on machined surfaces to smoothen the surface irregularities. The process results in improved surface finish, microhardness, resistance to wear and corrosion, fatigue life, and creep life. In the current work, the effect of ball burnishing process on the ultimate tensile strength (UTS) of Al7075-T6 alloy is analyzed using Taguchi method. The effect of four control factors, namely burnishing speed, burnishing feed, burnishing depth, and number of passes on the tensile strength, is studied by adopting L9 array; process parameters are optimized to fix the achievable maximum tensile strength for the said alloy. The results show that the burnishing process increased the tensile strength by 7% over the unburnished specimen.

**Keywords** Cold working · Finishing process · Ball burnishing · Al7075-T6 alloy · Tensile strength

## 1 Introduction

Removal of material as fine chips during conventional finishing processes such as grinding, lapping, honing, or super-finishing affects the service life because of the residual tensile stresses on the surface of the workpieces. This drawback can be overcome by using simple and effective finishing process which is called burnishing. It works on simple principle: When a metal surface undergoes plastic deformation, the surface irregularities diminish [there is no material removal] resulting in compressive residual stresses [14]. The burnishing process uses a hardball or roller that is highly finished to apply force on a machined surface and to flatten the peaks distributed on the work surface. This mechanism of filling the valleys with material from the peaks present over machined surface decreases the surface roughness. The compressive stresses induced at the surface during burnishing tend to increase microhardness. Usage of conventional machine tool with simple tooling makes the burnishing process

P. Kumara (✉) · U. P. Handadi
Shri Madhwa Vadiraja Institute of Technology and Management, Bantakal, Udupi 574115, India
e-mail: pavan.mech@sode-edu.in

 
P. Srinivasa Pai and V. Krishnaraj (eds.), *Sustainable Machining Strategies for Better Performance*, Lecture Notes in Mechanical Engineering,
https://doi.org/10.1007/978-981-16-2278-6_12

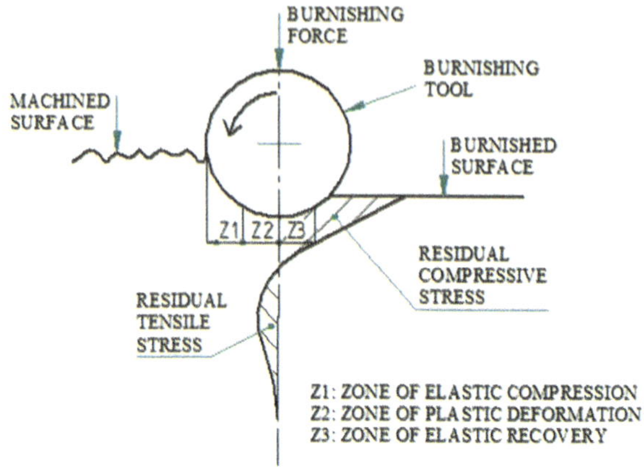


**Fig. 1** Schematic representation of the burnishing process

very much economical, resulting in the benefit of time and cost-saving. Figure 1 shows the burnishing mechanism and the stresses developed on the surface of a specimen.

Numerous studies have been done and are being undertaken in order to document the effect of the burnishing process: There were many objectives, however. Other than varying the type of materials, investigators have varied the tool parameters. The process came into existence as a sophisticated finishing operation during the late 1960s in metal industries. The different forms of burnishing—namely ball burnishing [9], roller burnishing, vibratory burnishing [16], centerless burnishing [3], laser-assisted burnishing [22]—are being tested for their performance even now. Various workpiece materials like brass [4], aluminum [6], Al–Cu alloys [10], mild steel [18], EN31 steel [1], titanium alloy [19], polymers [13] are burnished to test the effect of burnishing process on different kinds of materials. The results of these studies show that the burnishing process is capable of increasing various surface characteristics in each of these materials. To get the desired results during burnishing, it is necessary to use optimized control factors: In this context, many researchers have studied the influence of control factors during burnishing. The factors like burnishing force, burnishing speed, burnishing feed, number of passes [2, 11, 14], ball diameter [5], lubricant [12] are optimized with respect to few selected materials using various statistical tools. The findings can be applied in general to have better results in a given burnishing process.

The improvements in properties like surface finish [21], microhardness [20], wear resistance [8], corrosion resistance, fatigue strength [17], tensile strength are documented in various studies. It has been found that the burnishing process is capable of increasing ductility [7] of the workpiece material. Recent studies indicated that the burnishing process can be employed on implants to improve their corrosion resistance properties. In current work, the ball burnishing process is carried out on Al7075 T6 workpiece specimen samples to test its effect on the tensile strength of the material.

**Table 1** Chemical composition of Al 7075 T6 (by weight %)

| Si | Fe | Co | Mn | Mg | Cr | Zn | Other | Remd. (Al) |
|---|---|---|---|---|---|---|---|---|
| 0.05 | 0.309 | 1.09 | 0.025 | 1.61 | 0.105 | 4.22 | 0.0245 | 92.6 |

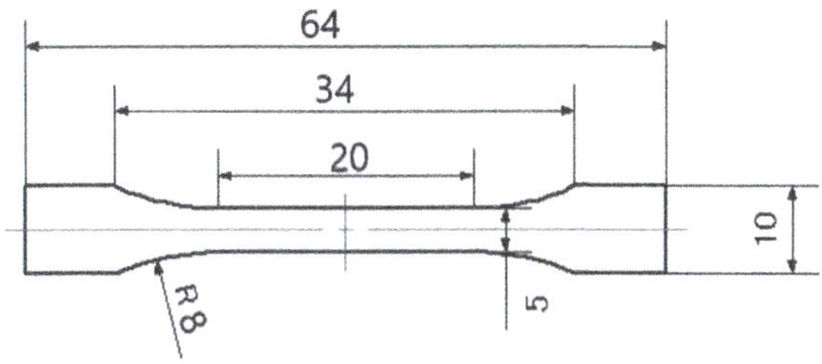


**Fig. 2** Specimen for burnishing test as per ASTM-E8 standards

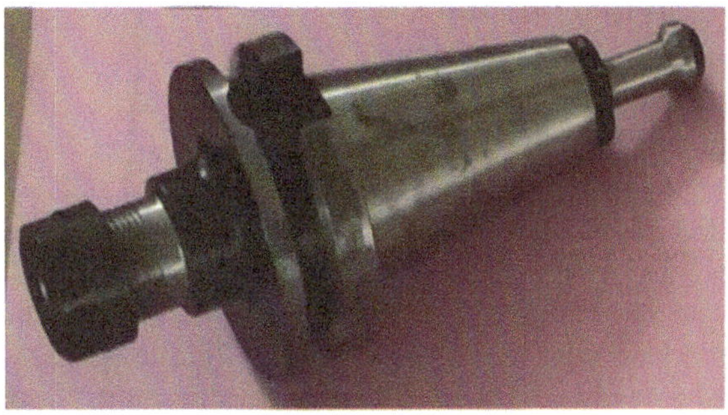

**Fig. 3** View of burnishing tool attachment

## 2 Workpiece Material Composition and Test Specimen Design

For the burnishing process, it was decided to select Al7075-T6 as test material because of its current use in various applications, particularly in the aerospace sector where strength/weight ratio should be high. The chemical composition of the workpiece material is given in Table 1. The workpiece material was received in sheet form of 350 × 250 × 1 mm dimensions. To get specimens for carrying out the burnishing tests, the sheets were cut to suitable dimensions as per ASTM-E8 standards—the dimensions are shown in Fig. 2.

## 3 Burnishing Ball and Attachment Tool

The burnishing process is carried out using a carbon–chromium ball of Ø8 mm as the burnishing element. The surface finish and hardness of the ball material are 0.012 μm and 65Hrc, respectively. For carrying out the process, a special tool holder

**Table 2** Process parameters and their respective levels

| Process parameters | Level 1 | Level 2 | Level 3 |
|---|---|---|---|
| Speed, rpm | 500 | 750 | 1000 |
| Feed, mm/rev | 0.05 | 0.1 | 0.15 |
| Depth of cut, mm | 0.5 | 0.7 | 0.1 |
| Number of passes (NOP) | 1 | 2 | 3 |

attachment is developed for use with the CNC vertical machining center (Make: AMS MCV350).

The tool holder is developed by modifying a standard one which can be accommodated in the machine easily—this is shown in Fig. 3.

## 4 Experimentation

Experimental works are carried out on CNC vertical milling center with the help of developed ball burnishing tool attachment. The parameters are set according to Taguchi design during each experimental run as shown in Table 2. During the burnishing process, a coolant is used in the burnishing zone to maintain the cold working conditions. Each specimen is marked to identify the burnishing regions as per calculated dimensions before carrying out the burnishing operations. Vertical milling machine used was AMS MCV 350. Tensile testing was done by using tensometer, adhering to ASTM E8 standards.

## 5 Design of Experiments

Taguchi L9 array orthogonal design methodology was adopted to design the experiments. Effect of four of the 'control factors,' each with three levels, on tensile strength values of Al7075-T6 specimens was investigated. The control factors and their levels are given in Table 3. Data on experimental runs according to Taguchi L9 array appear in Table 2. The S/N ratio of 'larger-the-better' characteristics is used for fixing the tensile strength values since the maximum tensile strength is desired in the work material which has undergone the burnishing process. MINITAB 17 software tool is used to analyze the results.

## 6 Results and Discussion

The tensile strength values of the unburnished specimen and those of burnished specimen of Al7075-T6 are given in Table 3. The maximum tensile strength obtained

**Table 3** Taguchi L9 array and tensile strength

| Run | Speed (rpm) | Feed (mm/rev) | Depth of cut (mm) | NOP | Tensile stress at max. load (MPa) |
|---|---|---|---|---|---|
| 1 | 500 (1) | 0.05 (1) | 0.5 (1) | 1 (1) | 587.84 |
| 2 | 500 (1) | 0.1 (2) | 0.7 (2) | 2 (2) | 550.27 |
| 3 | 500 (1) | 0.15 (3) | 0.1 (3) | 3 (3) | 579.70 |
| 4 | 750 (2) | 0.5 (1) | 0.7 (2) | 3 (3) | 602.50 |
| 5 | 750 (2) | 0.1 (2) | 0.1 (3) | 1 (1) | 611.47 |
| 6 | 750 (2) | 0.15 (3) | 0.5 (1) | 2 (2) | 568.98 |
| 7 | 1000 (3) | 0.05 (1) | 0.1 (3) | 2 (2) | 595.41 |
| 8 | 1000 (3) | 0.1 (2) | 0.5 (1) | 3 (3) | 599.77 |
| 9 | 1000 (3) | 0.15 (3) | 0.7 (2) | 1 (1) | 604.98 |
| – | Unburnished | | | | 569.98 |

was 611.47 MPa (experiment 5), while the lowest tensile strength was 550.27 MPa (experiment 2). The results indicated a maximum of 7% improvement in tensile strength values over the values assigned to the unburnished specimen.

## 6.1 *Analysis of Variance (ANOVA)*

The significance of each control factor on response is determined by using ANOVA technique, and the results are given in Table 4. The effect of burnishing parameters such as burnishing speed, feed, depth of cut, and the number of passes is calculated by using Minitab 17 software tool.

The number of burnishing passes is found to contribute to the tensile strength by 46.99% followed by speed with a contribution of 40.73%. The influence of feed and depth of cut is not found to be significant in improving tensile strength value of specimen. Figure 4 shows the contribution chart on tensile strength.

**Table 4** ANOVA for tensile strength

| Source | DoF | Adj.SS | Adj. Ms | *P*-value | % contribution on tensile strength |
|---|---|---|---|---|---|
| Speed | 2 | 1257.9 | 628.94 | 0.40 | 40.73 |
| Feed | 2 | 186.6 | 93.28 | 0.06 | 6.04 |
| Depth of cut | 2 | 192.4 | 96.22 | 0.06 | 6.23 |
| NOP | 2 | 1451.4 | 725.69 | 0.46 | 46.99 |
| Total | 8 | 3088.2 | | | |

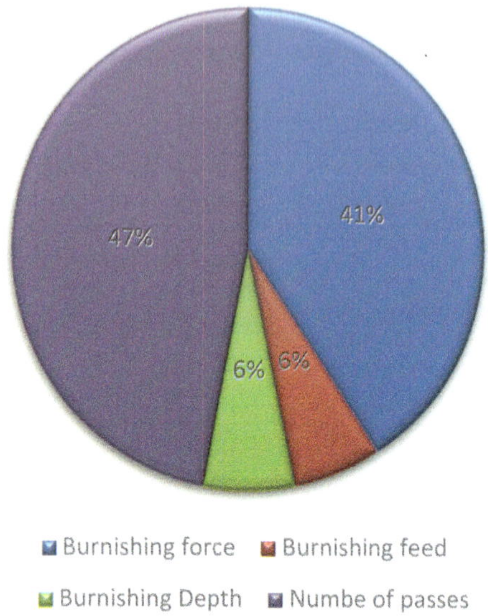


**Fig. 4** Percentage contribution of various burnishing factors on tensile strength values of Al7075 T6 specimen

## 6.2 *Signal-to-Noise Ratio*

Table 5 gives the ranks of factors with respect to their effect on tensile strength. Table indicates that the number of passes of the tool has the highest effect on tensile strength values followed by the effects of other three factors, namely speed, feed, and depth of cut, in the order. Therefore, it can be stated that by controlling the number of passes, one can alter the tensile strength value of the workpiece during the burnishing operation.

Figure 5 depicts the main effect plot of process parameters on tensile strength. The optimized parameter levels are speed 1000 rpm, feed 0.05 mm/rev, depth of cut 1 mm, and 1 number of pass of the tool. Tensile strength of the specimen increases with speed, depth of cut, and the number of passes; this is due to strengthening of the surface which occurs because of the cold working at higher speed levels.

The reason for reduction in strength with increasing feed is due to increase in temperature which results in softening of the surface. Figure 6 shows the results of

**Table 5** S/N ratio of tensile strength of burnished specimens

| Level | Speed | Feed | DOC | NOP |
|---|---|---|---|---|
| 1 | 55.15 | 55.49 | 55.35 | 55.58 |
| 2 | 55.48 | 55.37 | 55.35 | 55.14 |
| 3 | 55.56 | 55.33 | 55.50 | 55.47 |
| Delta | 0.41 | 0.16 | 0.15 | 0.45 |
| Rank | 2 | 3 | 4 | 1 |

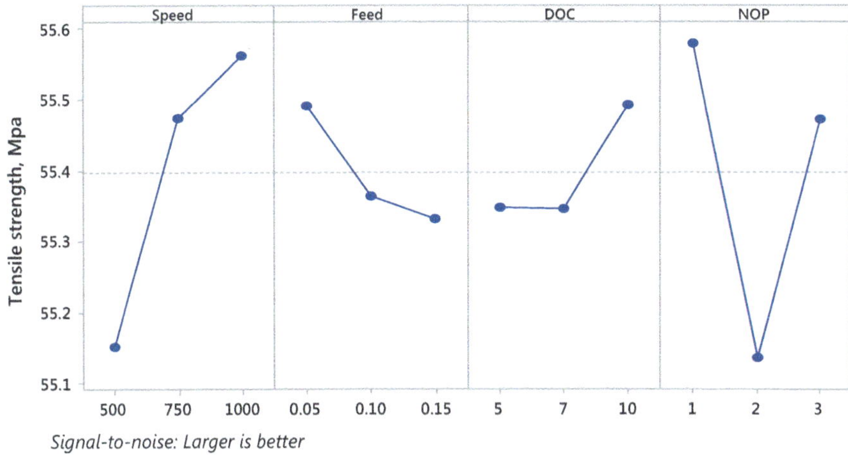


**Fig. 5** Effects plot for S/N ratio of tensile strength

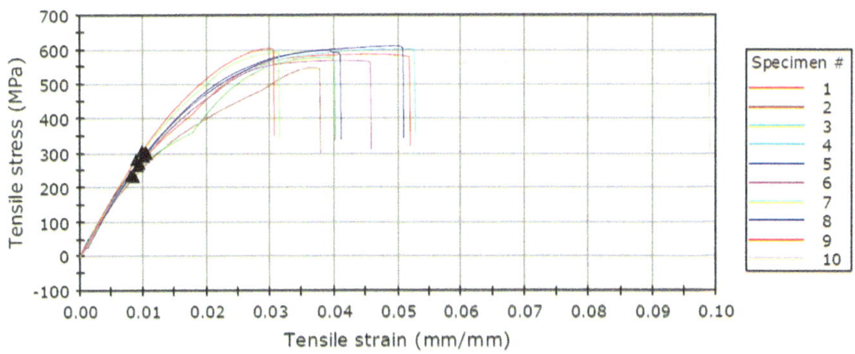


**Fig. 6** Stress–strain curve for burnished (1–9) and unburnished (10) specimen

tensile tests of burnished specimen along with unburnished one. The unburnished sample's tensile strength is lower than tensile strength of all burnished samples considered in the investigation which indicates that burnishing process is capable of improving the tensile strength of the specimen material.

## 7 Conclusion

The tensile strength of the Al7075 T6 material increases due to burnishing process carried out as per the present investigation. The selection of the range of process parameters is very crucial to get the desired results in a given burnishing process. The improvement in tensile strength value of the material selected can be achieved by adopting a higher speed, lower feed rate, higher depth of cut, and lower number of passes. The tensile strength of the said material after ball burnishing increased by 7% as compared to unburnished specimen.

## References


1. Babu PR (2016) The effect of external roller-burnishing on the surface roughness and surface hardness of EN 31 alloy steel. i-Manag J Future Eng Technol 4(1):36–41
2. Dabeer PS, Purohit GK (2010) Effect of ball burnishing parameters on surface roughness using surface roughness methodology. Adv Prod Eng Manage 5(2):111–116
3. El-Axir MH, Ibrahim AA (2005) Some surface characteristics due to center rest ball burnishing. J Mater Process Technol 167(1):47–53
4. El-Axir MH, Othman OM, Abodiena AM (2008) Improvements in out-of-roundness and microhardness of inner surfaces by internal ball burnishing process. J Mater Process Technol 196(1–3):120–128
5. El-Tayeb NSM, Low KO, Brevern PV (2009) On the surface and tribological characteristics of burnished cylindrical Al-6061. Tribol Int 42(2):320–326
6. Gharbi F, Sghaier S, Morel F, Benameur T (2015) Experimental investigation of the effect of burnishing force on service properties of AISI 1010 steel plates. J Mater Eng Perform 24(2):721–725
7. Gharbi F, Sghaier S, Hamdi H, Benameur T (2012) Ductility improvement of aluminum 1050A rolled sheet by a newly designed ball burnishing tool device. Int J Adv Manuf Technol 60(1–4):87–99
8. Hamadache H, Laouar L, Zeghib NE, Chaoui K (2006) Characteristics of Rb40 steel superficial layer under ball and roller burnishing. J Mater Process Technol 180(1–3):130–136
9. Hassan AM, Al-Jalil HF, Ebied AA (1998) Burnishing force and number of ball passes for the optimum surface finish of brass components. J Mater Process Technol 83:176–179
10. Hassan AM, Maqableh AM (2000) The effects of initial burnishing parameters on non-ferrous components. J Mater Process Technol 102(1):115–121
11. Ibrahim AA, Khalil T, Tawfeek T (2015) Study the influence of a new ball burnishing technique on the surface roughness of AISI 1018 low carbon steel. Int J Eng Technol 4(1):227–232
12. Loh NH, Tam SC, Miyazawa S (1989) A study of the effects of ball-burnishing parameters on surface roughness using factorial design. J Mech Working Technol 18(1):53–61
13. Low KO, Wong KJ (2011) Influence of ball burnishing on surface quality and tribological characteristics of polymers under dry sliding conditions. Tribol Int 44(2):144–153
14. Mahmood Hassan A, Al-Dhifi SZS (1999) Improvement in the wear resistance of brass components by the ball burnishing process. J Mater Process Technol 96(1–3):73–80
15. Jayakrishnan J, Suraj R (2014) Effect of roller burnishing process on tool steel material using CNC lathe. Int J Des Manuf Technol 5(3):155–159
16. Pande SS, Patel SM (1984) Investigations on vibratory burnishing process. Int J Mach Tool Des Res Pergamon 24(3):195–206
17. Prevéy P, Mahoney M (2003) Improved fatigue performance of friction stir welds with low plasticity burnishing : residual stress design and fatigue performance assessment. Mater Sci Forum 426:2933–2940
18. Rao MJ, Reddy CKA, Rao RP (2010) The effect of roller burnishing on surface hardness and surface roughness on mild steel specimens. Int J Appl Eng Res 1(4):777–785
19. Revankar GD, Shetty R, Rao SS, Gaitonde VN (2014) Analysis of surface roughness and hardness in ball burnishing of titanium alloy. Measurement 58:256–268
20. Seemikeri CYA, Brahmankar PK, Mahagaonkar SB (2008) Investigations on surface integrity of AISI 1045 using LPB tool. Tribol Int 41(8):724–734

21. Singh H, Kumar H, Parkash C (2015) Technology and research developments in magnetic abrasive finishing (MAF): a review. In: 5th International conference on Advances in Materials and Manufacturing Technology, Inproceedings, Chitkara University, Punjab, pp 38–40
22. Tian Y, Shin YC (2007) Laser-assisted burnishing of metals. Int J Mach Tools Manuf 47(1):14–22

# Comparison of GRA and TOPSIS Optimization Techniques in DMLS-Processed Bronze–Nickel Samples

**R. Rajesh, Mithun V. Kulkarni, P. Sampathkumaran, P. Sathish, and S. Sreenivas**

**Abstract** The article compares grey relational analysis (GRA) and technique for order preference by similarity to ideal solution (TOPSIS) methods for multi-objective optimization of direct metal laser sintering (DMLS) parameters to find the best set of parameters for the responses in the DMLS process. Taguchi design (two factors and three-level design) is used in this research. The factors include deposition thickness and scanning speed. The effects of these parameters are studied on the energy density, porosity, density and hardness, which are the responses. Analysis of variance (ANOVA) is used to find the most significant factor for the multi-responses. The comparison of the results suggests that both techniques produced a similar set of optimized parameters.

**Keywords** DMLS · GRA · TOPSIS · Energy density · Laser scan speed · Laser power

## 1 Introduction

It is known that almost all conventional manufacturing processes such as casting, forming, machining and welding are subtractive in nature as material removal takes place in these cases. Additive manufacturing (AM) being an innovative and a novel process, building up of material takes place layer by layer to the required thickness

R. Rajesh (✉)
Global Academy of Engineering, Bengaluru, India

M. V. Kulkarni
VijayaVittala Institute of Technology, Bengaluru, India

P. Sampathkumaran
Sambhram Institute of Technology, Bengaluru, India

P. Sathish
University of Technology and Applied Sciences, Salalah, Oman

S. Sreenivas
AMC College of Engineering, Bengaluru, India

 
P. Srinivasa Pai and V. Krishnaraj (eds.), *Sustainable Machining Strategies for Better Performance*, Lecture Notes in Mechanical Engineering,
https://doi.org/10.1007/978-981-16-2278-6_13

and hence the process has gained greater importance in these days. There are several additive manufacturing methods such as fused deposition modelling (FDM), laminated object manufacturing (LOM), stereolithography (SLA), selective laser melting (SLM), direct metal laser sintering (DMLS) and so on.

Among them, the DMLS process is quite popular and widely used. The process parameters in the DMLS technique has to be properly selected and controlled to produce engineering parts having minimal defects. It is reported in the literature that laser scan speed (LSS), laser power (LP), and laser scan spacing (LSSp) carried out using laser sintering technique [1] affect the properties (physical, metallurgical and mechanical) and in turn the quality of the parts. It is understood that LSS, LP, and LSSp play an important role as they are related directly to the amount of energy supplied for the sintering of the powder during processing. A low energy density provides low bonding and insufficient melting with the substrate; and samples so produced will not serve the purpose for further use. Hence, for material to perform better in service, it should possess good mechanical properties such as high hardness, good bond strength and fine-grained structure. Suarsana and Soenoko [2] reported that the increased holding time and sintering temperature may lead to higher hardness and better density due to the stronger or denser medium obtained during the process.

Kempen et al. [3] studied hardness and density with the change in layer thickness of 18Ni-300 steel. Density and hardness decreased with the increase in the layer thickness from 30 to 60 $\mu$m. From their investigation, they concluded that the increase in the layer thickness and scan speed were inversely proportional to the density and macrohardness. However, the change in microhardness was very minimal. It was also reported that the age hardening of the material has made good influence on the mechanical properties. Similarly, the work reported by Sun et al. [4] has used titanium alloy powder (Ti-6Al-4V) to produce test coupons made by the SLM technique. They reported that when the layer thickness and scan speed increases, the density decreases. In another work, Deffley [5] noted that the buildup of the thickness of layers in INCONEL 718 by the additive method reduced the density and in turn observed a reduction in the mechanical properties, and the reason for this has been pointed towards high porosity levels. Both the microstructure and mechanical properties were evaluated for a range of processing conditions and optimized.

Dingal et al. [6] in their work on the selective laser sintering process observed that by increasing the layer thickness, the porosity and density decreased and in turn the hardness. Further, it was reported that the porosity and density made a good influence on the layer thickness due to the balling effect, which may be attributed to the unfavourable defect acquired during the process. Liu et al. [7] mainly concentrated on the particle size distribution of stainless steel powders affecting the process parameters, viz. scanning speed and laser beam diameter; it is informed from this work that these process parameters have an effect on the density, ultimate tensile strength (UTS) and hardness. Increasing the scanning speed decreased the density.

Gu et al. [8] worked on stainless steel samples produced by the SLM process. They revealed that at higher scan speeds, the porosity level increases and the energy density decreases. Song et al. [9] observed a similar trend in an investigation, wherein it was noted that the microhardness decreased when the scanning speed increased from 100 to 300 mm/s at the horizontal surface for Ni–Cr alloy, whereas at the vertical surface, both hardness and tensile properties showed an increase thus revealing two different trends. These data trends have been explained based on the microstructural features.

Optimization of the DMLS process parameters is governed by its knack to control and measure the process variables involved in the DMLS process. The parameters which are of prime importance in the DMLS process include the scanning speed and material deposition thickness. Although there are many optimization techniques available, yet GRA and TOPSIS are widely used tools that help in transforming the multi-objective function into a single-objective function (SOF) [10]. Raju et al. [11] used the GRA technique for prototyping the rapid tooling parts using the SLA technique under the controlled machine parameter setting. Taguchi-based grey relational analysis was used to optimize the parameter levels. Anjian and Joseph [12] worked on optimizing the printing parameters that affect the surface roughness of the parts produced by the FDM process using Taguchi method. With optimized results, confirmatory printing was done and improved surface roughness was achieved.

Vishwas et al. [13] worked on materials such as ABS and nylon using the FDM technique to assess the properties like the dimensional accuracy and the UTS. Taguchi methodology was used. Uzair et al. [14] studied the impact of the fused deposition modeling (FDM) process parameters on the strength of built parts using Taguchi's design of experiments.

It is thus realized from the works of the literature that the GRA and TOPSIS techniques have not been used in optimizing the DMLS parameters. Thus, research compares and finds the optimized DMLS process parameters using the above-said techniques.

## 2 Materials and Methods

### *2.1 Materials*

Bronze–nickel (DirectMetal 20) powder is a very fine-grained powder that results in components with excellent mechanical properties, surface quality and accuracy. The specialty of this powder mixture is that it partially compensates for the natural sintering shrinkage occurring due to the laser sintering process. Typical applications include injection moulding tooling's, direct manufacture of functional metal prototypes, and also, the parts built using this powder mixture have very good corrosion resistance [15]. For preparing the samples, the powder was procured from EOS GmbH, Germany, of mesh size 20 $\mu$m. The nominal composition of the powder is shown in Table 1.

**Table 1** Nominal powder composition

| Material | Composition (wt. %) | | | |
|---|---|---|---|---|
| | Cu | Ni | Sn | P |
| Bronze–nickel | 76.6 | 14.8 | 7.0 | 1.6 |

**Fig. 1** DMLS processed samples

## 2.2 *Specimen Buildup by DMLS Process*

In the first instance, 3D CAD models of test specimens were modelled using CATIA R19 V5 CAD/CAM software. They were then converted into .stl file and then converted to .bdf format using EOS RP tool software. The models were sliced into layers using software with the layer thickness chosen to be 20, 40 and 60 μm. The specimens of bronze–nickel parts with a cylindrical shape of diameter 10 mm and thickness 8 mm were built as per our convenience (Fig. 1). The DMLS samples were processed in an EOSINT M250 Xtended RP machine of the base plate of size 250 mm × 250 mm × 25 mm acting as a build platform. The powder was preheated to avoid the build failure. Studies have shown that preheating improves the mechanical strength of the sintered powder, reduces warping and the in-built residual stresses of AM components [16]. The microwire electrodischarge method was used to remove the samples from base plate.

## 2.3 *Energy Density*

Laser power decides the amount of melting the powder can undergo, or in other words, laser power defines the severity of the temperature gradient in the powder, while the amount of energy input during the melting defines the term scan speed. Both, the laser power and scan speed, have a direct effect on the surface quality. During laser scanning, the distance between the central lines of two neighbouring scanning paths becomes important, and thus, the term hatch spacing is used. It must be noted that poor hatch spacing often results in regular porosity in built parts as adjacent melt lines do not fuse together completely [17]. Thus, the energy density was calculated using Eq. 1, which relates the energy density with LP, LSS and layer thickness [12].

$$\mathrm{ED} = \frac{\mathrm{N}}{\mathrm{v} \times \mathrm{h} \times \mathrm{t}} \tag{1}$$

where ED = energy density ($J/mm^3$), $N$ = LP (W), $v$ = LSS (mm/s), $h$ = hatch spacing (mm) and $t$ = layer thickness (mm).

## 2.4 *Characterization Methods*

**Density**

The density measurements were carried in accordance to the Archimedes principle.

**Porosity**

The theoretical density of the specimens and the porosity in the test specimens were determined using Eqs. 2 and 3.

$$\rho_t = \frac{100}{\sum_{i=1}^{n} \frac{x_i}{\rho_i}} \tag{2}$$

where $\rho_t$ being the theoretical density, $x_i$ is the fraction of component mixture of all material ($i$ being Cu, Ni, P and Sn), $\rho_i$ is each components density ($g/cm^3$) and $\rho_e$ is the experimental density ($g/cm^3$).

$$\% \text{ of porosity} = \frac{\rho_t - \rho_e}{\rho_t} \times 100 \tag{3}$$

**Macrohardness**

One of the most commonly tested mechanical properties is the hardness. Hardness testing is relatively easier and inexpensive and provides an insight to properties such as wear resistance and yield strength. Due to the presence of internal defects, AM components exhibit variations in yield strength when tested for repeatability, but often it is seen that hardness data are unaffected by the internal defects. Thus, the hardness measurements indicate the true effect of microstructural features. The Vickers hardness test is a versatile hardness test method used for both macro- and microhardness testing. It has a broad load range and is suitable for a wide range of applications and materials. The Vickers hardness test is often regarded as easier to use than other hardness tests [18]. Vickers macro-indentation hardness (VHN) of AM samples is subjected to a load of 15 kg. The load is normally applied for 10–15 s using a square base diamond indenter. The diagonal of the indentation is measured after the removal of the load using a travelling microscope. Equation 3 was used in the calculation of the hardness.

$$VHN = 1.854\frac{W}{d^2} \tag{4}$$

where ‘$W$’ is the load applied in kg and ‘$d$’ is the arithmetic mean of the two diagonals in mm.

## 2.5 Taguchi Method

Based on the number of input parameters and their levels involved in the investigation, an appropriate orthogonal array (OA) needs to be selected. Depending on the number of parameters and their levels (Table 2), the minimum number of experimental runs required for investigation is obtained from the relation $(L - 1)P + 1$, where $L$ is the number of levels and $P$ is the number of input parameters [19]. In the present study, since $L = 3$, $P = 2$, the minimum number of experiments required is 5. But, we adopted L9 design for experimentation. The L9 orthogonal array with input parameters and responses [energy density (ED), porosity ($P$), hardness ($H$) and density ($\rho_e$)] measured is indicated in Table 3 designed using MINITAB V16 software.

**Table 2** DMLS parameters and their levels

| Input parameters | Level 1 | Level 2 | Level 3 |
|---|---|---|---|
| Laser scan speed (LSS), mm/s | 100 | 200 | 300 |
| Layer thickness (TM), mm | 0.02 | 0.04 | 0.06 |

**Table 3** L9 Orthogonal array with input parameters and output responses

| Exp. no | Input parameters | | Responses | | | |
|---|---|---|---|---|---|---|
| | LSS (mm/s) | TM (mm) | ED (J/mm$^3$) | $P$ (%) | $\rho_e$ (kg/cm$^3$) | $H$ (VHN) |
| 1 | 100 | 0.02 | 600 | 3.722 | 8.02 | 146 |
| 2 | 100 | 0.04 | 300 | 4.802 | 7.93 | 95.8 |
| 3 | 100 | 0.06 | 200 | 5.042 | 7.91 | 90.8 |
| 4 | 200 | 0.02 | 300 | 6.363 | 7.8 | 111.2 |
| 5 | 200 | 0.04 | 150 | 6.963 | 7.75 | 78.4 |
| 6 | 200 | 0.06 | 100 | 7.083 | 7.61 | 74.8 |
| 7 | 300 | 0.02 | 200 | 6.483 | 7.79 | 78.9 |
| 8 | 300 | 0.04 | 100 | 7.083 | 7.74 | 66.4 |
| 9 | 300 | 0.06 | 66.66 | 7.203 | 7.31 | 65.2 |

# 3 Results and Discussion

## 3.1 Grey Relational Approach (GRA)

GRA is a method of transforming a multi-response problem into a single-objective problem. Deng in 1982 developed the GRA technique. Grey relational grade (GRG) is used in assessing the overall performance of the experimental tests. The optimal solution using GRA is determined by ranking the GRG values. The highest GRG value is ranked 1. The process of performing GRA has been explained in the following sections.

**Stage 1**

The S/N ratio has been used to assess the obtained results. It is used to know the amount of valuable information from the given data set. For the output responses, S/N ratios were determined by using Eqs. 5 and 6. The ratios can be written off as (i) smaller-the-better (SB), (ii) nominal-the-better (NB) and (iii) larger-the-better (LB). In the present work, the 'LB option' has been used for the responses (ED, $\rho_e$ and H) and the 'SB option' has been used for 'porosity', as it is always desired to keep the level of porosity to a minimum. The calculated S/N ratios are presented in Table 4.

1. Larger-the-better

$$\text{S/N ratio} = -10\log\left(\frac{1}{n}\right)\sum_{i=1}^{n}\frac{1}{y_{ij}^2} \tag{5}$$

2. Smaller-the-better

.

Table 4 S/N ratio for the output responses

| Experiment no | S/N ratio | | | |
|---|---|---|---|---|
| | ED (LB) | *P* (SB) | $\rho_e$(LB) | *D* (LB) |
| 1 | 55.563 | −11.416 | 18.083 | 43.287 |
| 2 | 49.542 | −13.628 | 17.985 | 39.627 |
| 3 | 46.021 | −14.052 | 17.964 | 39.162 |
| 4 | 49.542 | −16.073 | 17.842 | 40.922 |
| 5 | 43.522 | −16.856 | 17.786 | 37.886 |
| 6 | 40.000 | −17.004 | 17.628 | 37.478 |
| 7 | 46.021 | −16.236 | 17.831 | 37.942 |
| 8 | 40.000 | −17.004 | 17.775 | 36.443 |
| 9 | 36.477 | −17.150 | 17.278 | 36.285 |

$$\text{S/N ratio} = -10\log\left(\frac{1}{n}\right)\sum_{i=1}^{n} y_{ij}^2 \tag{6}$$

where '$n$' represents the number of repetitions for each experiment and $y_{ij}$ are response values.

**Stage 2**

The S/N ratio values obtained from stage 1 were normalized using Eqs. 7 and 8. Normalization involves transforming the S/N ratio values into a dimensionless parameter and setting the obtained values in the range of zero to one.

1. Larger-the-better

$$N_i^*(k) = \frac{y_i(k) - \min\ y_i(k)}{\max\ y_i(k) - \min\ y_i(k)} \tag{7}$$

2. Smaller-the-better

$$N_i^*(k) = \frac{\max y_i(k) - y_i(k)}{\max y_i(k) - \min y_i(k)} \tag{8}$$

where $i = 1, 2, \ldots, m$ and $m =$ the number of trial data, $k$ is the response value ($k = 1, 2, 3, \ldots, n$), $n =$ the number of factors, $y_i(k) =$ original sequence, $N_i^*(k)$ is the normalized value, min $y_i(k)$ is the minimum of $y_i(k)$ and max $y_i(k)$ is the maximum value of $y_i(k)$. Table 5 presents the normalized values.

**Table 5** Normalized value for the output responses

| Experiment no | Normalized value | | | |
|---|---|---|---|---|
| | ED (LB) | $P$ (SB) | $\rho_e$ (LB) | $D$ (LB) |
| 1 | 1.000 | 0.000 | 1.000 | 1.000 |
| 2 | 0.685 | 0.386 | 0.878 | 0.477 |
| 3 | 0.500 | 0.460 | 0.851 | 0.411 |
| 4 | 0.685 | 0.812 | 0.700 | 0.662 |
| 5 | 0.369 | 0.949 | 0.631 | 0.229 |
| 6 | 0.185 | 0.975 | 0.434 | 0.170 |
| 7 | 0.500 | 0.840 | 0.686 | 0.237 |
| 8 | 0.185 | 0.975 | 0.617 | 0.023 |
| 9 | 0.000 | 1.000 | 0.000 | 0.000 |

**Table 6** Deviation sequence, GRC, GRG and rank for all the output responses for bronze–nickel samples

| Exp. no | $\Delta_{oi}$ | | | | $\epsilon_i(k)$ | | | | GRG | Rank |
|---|---|---|---|---|---|---|---|---|---|---|
| | ED | $P$ | $\rho_e$ | $H$ | ED | $P$ | $\rho_e$ | $H$ | | |
| 1 | 0.000 | 1.000 | 0.000 | 0.000 | 1.000 | 0.333 | 1.000 | 1.000 | 0.833 | 1 |
| 2 | 0.315 | 0.614 | 0.122 | 0.523 | 0.613 | 0.449 | 0.804 | 0.489 | 0.589 | 3 |
| 3 | 0.500 | 0.540 | 0.149 | 0.589 | 0.500 | 0.481 | 0.770 | 0.459 | 0.553 | 7 |
| 4 | 0.315 | 0.188 | 0.300 | 0.338 | 0.613 | 0.727 | 0.625 | 0.597 | 0.640 | 2 |
| 5 | 0.631 | 0.051 | 0.369 | 0.771 | 0.442 | 0.907 | 0.575 | 0.393 | 0.579 | 4 |
| 6 | 0.815 | 0.025 | 0.566 | 0.830 | 0.380 | 0.952 | 0.469 | 0.376 | 0.544 | 8 |
| 7 | 0.500 | 0.160 | 0.314 | 0.763 | 0.500 | 0.758 | 0.614 | 0.396 | 0.567 | 5 |
| 8 | 0.815 | 0.025 | 0.383 | 0.977 | 0.380 | 0.952 | 0.566 | 0.338 | 0.559 | 6 |
| 9 | 1.000 | 0.000 | 1.000 | 1.000 | 0.333 | 1.000 | 0.333 | 0.333 | 0.500 | 9 |
| **Average of GRG** | | | | | | | | | **0.596** | |

**Stage 3**

The next step in the GRA process is finding the deviation sequence values $\Delta_{oi}(k)$. Normalized values in Table 5 were used in Eq. 9 to obtain the $\Delta_{oi}(k)$ values.

$$\Delta_{oi}(k) = \|y_o(k) - y_i(k)\| \tag{9}$$

where $y_o(k)$ and $y_i(k)$ are the reference and comparability sequences. The value of $y_o(k)$ is taken as 1 (one), and the values of $\Delta_{oi}(k)$ are listed in Table 6.

**Stage 4**

This stage involves finding of the grey relation coefficient (GRC) using the values obtained from the previous stage. GRC helps in finding the link between the ideal and actual normalized experimental results. Equation 10 was used to compute the GRC $\epsilon_i(k)$.

$$\epsilon_i(k) = \frac{\Delta\min + \tau\Delta\max}{\Delta_{oi}(k) + \tau\Delta\max} \tag{10}$$

where $\Delta_{\max}$ and $\Delta_{\min}$ are the highest and the least values of $\Delta_{oi}(k)$ and $\tau$ is taken as 0.5.

**Stage 5**

The grey relational grade (GRG) refers to the average of GRC of responses. GRG was calculated using Eq. 11.

**Table 7** Response table for the average GRGs

| Level | LSS | TM |
|---|---|---|
| 1 | **0.6583** | **0.6800** |
| 2 | 0.5877 | 0.5757 |
| 3 | 0.5420 | 0.5323 |
| Delta | 0.1163 | 0.1477 |
| Rank | 2 | 1 |

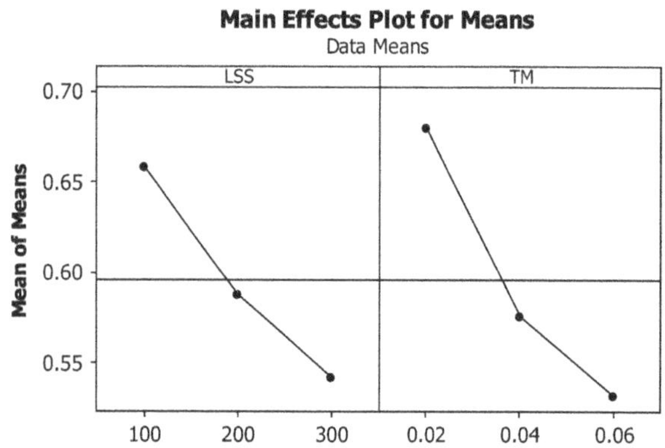


**Fig. 2** Main effects plot for means of GRG

$$\text{GRG} = \left(\frac{1}{m}\right) \sum \epsilon_i(k) \tag{11}$$

where '$m$' denotes the number of response variables.

If the GRG value of an experiment is one or is closer to one, then the GRG value corresponding to that particular experiment indicates the closeness to the optimal response. Further, GRG values are ranked as indicated in Table 6. It is seen that experiment 1 has the highest GRG value meaning very much closer or equal to one. However, in order to make sure that the highest-ranked experiment is the best optimal solution, the response table was drawn. Table 7 indicates the response table for average GRGs as obtained from Table 6. The layer thickness (TM) was found to be having a higher delta value than the LSS. Thus, from Fig. 2, it can be said that TM is the most influential factor and Level 1 of LSS and TM gives the optimized set of input parameters. Also, it is understood that the optimized set of parameters lies within the Taguchi's L9 experimentation model.

## 3.2 *Analysis of Variance (ANOVA)*

ANOVA of GRG is shown in Table 8. ANOVA helps to determine the percentage contribution of each input factor, or in other words, it helps to find the significance

**Table 8** ANOVA using GRG values in GRA

| Source | Degrees of freedom | Sum of squares | Mean square | $F$ value | Percentage contribution |
|---|---|---|---|---|---|
| LSS | 2 | 0.020613 | 0.010306 | 2.14 | 27.69 |
| TM | 2 | 0.034569 | 0.017284 | 3.59 | 46.44 |
| Error | 4 | 0.019241 | 0.004810 | | |
| Total | 8 | 0.074422 | | | |

level of individual parameters that affect the multiple responses of the process under investigation. Minitab 16 software was used to extract the values in the ANOVA table. Between the two input parameters of the DMLS process, TM has the most influence with a percentage contribution of 46.44%.

### *3.3 Technique for Order Preference by Similarity to Ideal Solution Method (TOPSIS)*

TOPSIS is a robust and widely accepted multi-criteria decision making (MCDM) technique in operation research and production engineering. In this method, the best alternatives are searched based on the closeness coefficient from the ideal best/worst solution. The basics behind the method are to evaluate the alternatives on the Euclidian distance scale such that the least span from ideal best and farthest from ideal worst solution is achieved. The alternatives are then ordered based on their rank.

**Stage 1**

Normalization of data was done using Eq. 12, and values are presented in Table 9.

**Table 9** Normalized and weighted normalized values for the output responses

| Exp. no | $R_{ij}$ | | | | $V_{ij}$ | | | |
|---|---|---|---|---|---|---|---|---|
| | ED | $P$ | $\rho_e$ | $H$ | ED | $P$ | $\rho_e$ | $H$ |
| 1 | 0.735 | 0.200 | 0.344 | 0.524 | 0.184 | 0.050 | 0.086 | 0.131 |
| 2 | 0.367 | 0.258 | 0.340 | 0.344 | 0.092 | 0.065 | 0.085 | 0.086 |
| 3 | 0.245 | 0.271 | 0.340 | 0.326 | 0.061 | 0.068 | 0.085 | 0.081 |
| 4 | 0.367 | 0.342 | 0.335 | 0.399 | 0.092 | 0.086 | 0.084 | 0.100 |
| 5 | 0.184 | 0.375 | 0.333 | 0.281 | 0.046 | 0.094 | 0.083 | 0.070 |
| 6 | 0.122 | 0.381 | 0.327 | 0.268 | 0.031 | 0.095 | 0.082 | 0.067 |
| 7 | 0.245 | 0.349 | 0.334 | 0.283 | 0.061 | 0.087 | 0.084 | 0.071 |
| 8 | 0.122 | 0.381 | 0.332 | 0.238 | 0.031 | 0.095 | 0.083 | 0.060 |
| 9 | 0.082 | 0.387 | 0.314 | 0.234 | 0.020 | 0.097 | 0.078 | 0.058 |

$$R_{ij} = \frac{X_{ij}}{\sqrt{\sum_{i=1}^{m} X_{ij}^2}}; \text{ for } i = 1, 2, \ldots, m;\ j = 1, 2, \ldots, n \tag{12}$$

where

$i$ number of alternatives (experimental runs).
$j$ number of criteria (output responses).
$X_{ij}$ normalized value of $i$th experimental run associated with $j$th output response.

**Stage 2**

Considering the relative importance weights ($Wj$) of different attributes with respect to the objective, the summation of weights should be equal to one. In this work there are four responses, and hence, 25% weightage was given to the each response. Thus, $Wj = 0.25$.

**Stage 3**

The weighted normalized matrix was constructed by taking the product of respective weights and the normalized values of the particular column (Eq. 13). The weighted normalized values ($V_{ij}$) are shown in Table 9.

$$V_{ij} = W_j R_{ij} \tag{13}$$

**Stage 4**

Using Eq. 14a and b, ideal and negative ideal solutions were calculated and the values are shown in Table 10. Positive and negative ideal solutions are given by:

$$A^{+} = \{V_1^{+} \ldots V_n^{+}\} \tag{14a}$$

$$A^{-} = \{V_1^{-} \ldots V_n^{-}\} \tag{14b}$$

where $V_j^{+} = \{\max(V_{ij}) \text{if } j \in J; \min(V_{ij}) \text{if } j \in J'\}$

$$V_j^{-} = \{\min(V_{ij}) \text{if } j \in J; \min(V_{ij}) \text{if } j \in J'\}$$

**Table 10** Positive and negative ideal solutions for output responses

| | ED | $P$ | $\rho_e$ | $H$ |
|---|---|---|---|---|
| A+ | 0.184 | 0.050 | 0.086 | 0.131 |
| A− | 0.020 | 0.097 | 0.078 | 0.058 |

**Stage 5**

The separation measure (SM) values, namely the ideal alternative ($S_i^+$) and the negative alternatives ($S_i^-$) were calculated using Eq. 15a and 15b for the values obtained in Table 10. The obtained SM values have been shown in Table 11.

$$S_i^+ = \sqrt{\sum_{(i=1)}^{n} \left[V_j^+ - V_{ij}\right]^2} \quad i = 1, 2 \ldots m \tag{15a}$$

$$S_i^- = \sqrt{\sum_{(i=1)}^{n} \left[V_j^- - V_{ij}\right]^2} \quad i = 1, 2 \ldots m \tag{15b}$$

**Stage 6**

Using Eq. 16, the relative closeness coefficients were calculated and shown as seen in Table 11. The option with Pi closest to 1 was selected.

$$P_i = \frac{S_i^-}{\left(S_i^+ + S_i^-\right)} 0 < P_i < 1 \tag{16}$$

Table 11 displays that among the nine experiments; trial 1 has the maximum relative closeness value. The higher the relative closeness value of an experiment, the higher is the chance of that experiment being closer to the ideal value. A response table has been constructed using $P_i$ values from Table 11 in order to know the optimized set of parameters for the TOPSIS technique. Table 12 displays the response table for the average grey relational grade, and thus, it is understood that the optimized set of parameters lies within the Taguchi's L9 experimentation model. The LSS was found

**Table 11** Separation measure, relative closeness value and ranking for the various experimental run

| Experiment no. | $S_i^+$ | $S_i^-$ | $P_i$ | Rank |
|---|---|---|---|---|
| 1 | 0.000 | 0.185 | 1.000 | 1 |
| 2 | 0.103 | 0.083 | 0.446 | 3 |
| 3 | 0.133 | 0.055 | 0.294 | 4 |
| 4 | 0.103 | 0.083 | 0.447 | 2 |
| 5 | 0.157 | 0.029 | 0.155 | 6 |
| 6 | 0.172 | 0.014 | 0.074 | 7 |
| 7 | 0.141 | 0.044 | 0.237 | 5 |
| 8 | 0.175 | 0.011 | 0.061 | 8 |
| 9 | 0.185 | 0.000 | 0.000 | 9 |
| **Average of relative closeness value** | | | **0.302** | |

**Table 12** Response table for the average closeness value

| Level | LSS | TM |
|---|---|---|
| 1 | **0.58014** | **0.56136** |
| 2 | 0.22532 | 0.22076 |
| 3 | 0.09947 | 0.12280 |
| Delta | 0.48067 | 0.43856 |
| Rank | 1 | 2 |

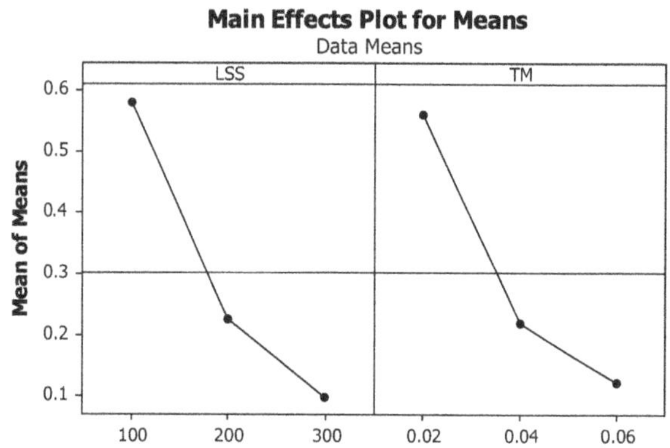


**Fig. 3** Main effects plot for means of average closeness value

to be having a higher delta value than the TM. Thus, from Fig. 3, it can be said that LSS as the most influential factor and Level 1 of LSS and TM gives the optimized set of input parameters.

## 3.4 *Analysis of Variance (ANOVA)*

ANOVA of the relative closeness value is shown in Table 13. Among the two input parameters of the DMLS process, the LSS has higher influence with a percentage contribution of 49.31% as against the layer thickness with a percentage contribution of 42.05%.

**Table 13** Analysis of variance (ANOVA) using the relative closeness values

| Source | Degrees of freedom | Sum of squares | Mean square | *F* value | Percentage contribution |
|---|---|---|---|---|---|
| LSS | 2 | 0.37279 | 0.18639 | 11.43 | 49.31 |
| TM | 2 | 0.31794 | 0.15897 | 9.74 | 42.05 |
| Error | 4 | 0.06525 | 0.01631 | | |
| Total | 8 | 0.75598 | | | |

**Table 14** Optimized and predicted GRG values of GRA and TOPSIS methods

| Methods | Optimized GRG value | Predicted GRG value | % Difference |
|---|---|---|---|
| GRA | 0.833 | 0.7423 | 10.88 |
| TOPSIS | 1 | 0.8395 | 16.05 |

### 3.5 Comparison of GRA and TOPSIS Results

Based on the study conducted, both the GRA and TOPSIS have produced results, within the Taguchi's L9 experimentation matrix. However, both methods have arrived at a different conclusion. GRA study says that TM has more influence over the process, whereas TOPSIS says that LSS has more influence over the process. Thus, in order to select the best method among the two, a comparison of the predicted closeness value ($\alpha$) was done using Eq. 17.

$$\alpha = \alpha_m + \sum (\alpha_n - \alpha_m) \tag{17}$$

where $\alpha_m$ = mean of the GRG values in the case of GRA and mean of the relative closeness value in the case of TOPSIS.

$\alpha_n$ = mean grey relational at the optimum level in the case of GRA and the average closeness value in the case of TOPSIS.

From Tables 6 and 11, the values of $\alpha_m(\text{GRA}) = 0.596$ and $\alpha_m(\text{TOPSIS}) = 0.302$. From Table 7 for GRA, $\sum(\alpha_n - \alpha_m) = \{(0.6583 - 0.596) + (0.6800 - 0.596)\} = 0.1463$, and from Table 12 for TOPSIS, $\sum(\alpha_n - \alpha_m) = \{(0.58014 - 0.302) + (0.56136 - 0.302)\} = 0.5375$. Thus, $\alpha(\text{GRA}) = 0.7423$and$\alpha(\text{TOPSIS}) = 0.8395$. Comparison ofoptimized and predicted GRG vales for both methods is shown in Table 14.

## 4 Conclusion

GRA and TOPSIS methods were successfully used for comparison and assessment of optimized values. The following conclusions are drawn

- Both in GRA and TOPSIS methods, the optimized set of parameters was found to be within the L9 experimental run.
- Experiment 1 gave the optimized set of parameters with LSS of 100 mm/s, TM of 0.02 mm, ED of 600 J/mm$^3$, porosity of 3.722%, density equal to 8.02 kg/cm$^3$, hardness equal to 146 VHN.
- Although it was clear from the experimental results that the first experimental trial gives the optimized set of parameters, yet GRA and TOPSIS methods were carried out to known if any other optimized set of parameters were available within the L9 experimentation run.

- For GRG value, ANOVA indicated that layer thickness was the major influencing factor on the overall objective, and in case of TOPSIS, laser scan speed was found to be the most influencing factor. However, since the percentage difference between the experimental and predicted values in case of GRA is less than TOPSIS method, we can conclude that layer thickness to be the most influencing factor in the DMLS process.

**Acknowledgements** Thanks to Dr. S. Sivamani and Mr. M Vijayanad, Lecturers, Engineering Department, University of Technology and Applied Sciences - Salalah, Oman, for proofreading the manuscript.

**Declaration of Competing Interest** The authors declare that they have no known competing financial interests or personal relationships that could have appeared to influence the work reported in this paper.

## References

1. Williams JD, Deckard CR (2011) Emerald article: Advances in modeling the effects of selected parameters on the SLS process. Rapid Prototyping J 4(2):90–100
2. Suarsana K, Soenoko R (2015) Hardness, density and porosity of Al/(SiCw + $Al_2O_3$p) composite by powder metallurgy process without and with sintering. Appl Mech Mater 776:246–252. https://doi.org/10.4028/www.scientific.net/amm.776.246
3. Kempen K, Yasa E, Thijs L, Kruth JP, Van Humbeeck J (2011) Microstructure and mechanical properties of selective laser melted 18Ni-300 steel. Phys Procedia 12(PART 1):255–263. https://doi.org/10.1016/j.phpro.2011.03.033
4. Sun J, Yang Y, Wang D (2013) Parametric optimization of selective laser melting for forming Ti6Al4V samples by Taguchi method. Opt Laser Technol 49:118–124. https://doi.org/10.1016/j.optlastec.2012.12.002
5. James DR (2012) Development of processing strategies for the additive layer manufacture of aerospace components in Inconel 718. University of Sheffield
6. Dingal S, Pradhan TR, Sundar JKS, Choudhury AR, Roy SK (2008) The application of Taguchi's method in the experimental investigation of the laser sintering process. Int J Adv Manuf Technol 38(9–10):904–914. https://doi.org/10.1007/s00170-007-1154-1
7. Liu B, Wildman R, Tuck C, Ashcroft I, Hague R (2011) Investigation the effect of particle size distribution on processing parameters optimisation in selective laser melting process. In: 22nd annual international solid freeform fabrication symposium—an additive manufacturing conference SFF 2011, no. January 2015, pp 227–238
8. Gu H, Gong H, Pal D, Rafi K, Starr T, Stucker B (2013) Influences of energy density on porosity and microstructure of selective laser melted 17-4PH stainless steel. In: 24th International SFF symposium—an additive manufacturing conference SFF 2013, no. March 2018, pp 474–489
9. Song B, Dong S, Coddet P, Liao H, Coddet C (2014) Fabrication of NiCr alloy parts by selective laser melting: columnar microstructure and anisotropic mechanical behavior. Mater Des 53:1–7. https://doi.org/10.1016/j.matdes.2013.07.010
10. Srinivasan L, Mohammad Chand K, Deepan Bharathi Kannan T, Sathiya P, Biju S (2018) Application of GRA and TOPSIS optimization techniques in GTA welding of 15CDV6 aerospace material. Trans Indian Inst Met 71(2):373–382. https://doi.org/10.1007/s12666-017-1166-y
11. Raju BS, Sekhar UC, Drakshayani DN (2014) Optimizing multiple quality characteristics of Stereolithography process via Taguchi method-based Grey Analysis for SL5530 epoxy resin

material to enhance part quality. Procedia Mater Sci 5:2532–2541. https://doi.org/10.1016/j.mspro.2014.07.505
12. Chen A, Chen JC (2018) Optimization of surface roughness in additive manufacturing processes via Taguchi methodology. Int J Ind Manuf Eng 12(4):370–374 [Online]. Available https://waset.org/publications/10008825/optimization-of-surface-roughness-in-additive-manufacturing-processes-via-taguchi-methodology
13. Vishwas M, Basavaraj CK, Vinyas M (2018) Experimental investigation using Taguchi method to optimize process parameters of fused deposition modeling for ABS and nylon materials. Mater Today Proc 5(2):7106–7114. https://doi.org/10.1016/j.matpr.2017.11.375
14. uz Zaman UK, Boesch E, Siadat A, Rivette M, Baqai AA (2019) Impact of fused deposition modeling (FDM) process parameters on strength of built parts using Taguchi's design of experiments. Int J Adv Manuf Technol 101(5–8):1215–1226. https://doi.org/10.1007/s00170-018-3014-6
15. EOS GmbH—Electro Optical Systems (2004) Direct Metal and Direct Steel materials for EOSINT M 250 Xtended, vol 49, pp 1–6 [Online]. Available https://www.3axis.us/matetials/dmls/DirectMetal20.pdf
16. Leung CLA, Tosi R, Muzangaza E, Nonni S, Withers PJ, Lee PD (2019) Effect of preheating on the thermal, microstructural and mechanical properties of selective electron beam melted Ti-6Al-4V components. Mater Des 174:107792. https://doi.org/10.1016/j.matdes.2019.107792
17. Yap et al CY (2015) Review of selective laser melting: materials and applications. Appl Phys Rev 2(4). https://doi.org/10.1063/1.4935926
18. Microhardness Testing. https://www.nanoscience.com/techniques/mechanical-testing/microhardness-testing/. Accessed 20 Nov 2020
19. Talib F, Rahman Z (2008) A study of optimization of process by using Taguchi's parameter design approach. IUP J Oper Manag, no. July 2015, pp 5–18

# Optimization of Parameters for Material Removal Rate and Surface Roughness in Wire Electric Discharge Grinding (WEDG) for Micro-machining of Cemented Carbide Rods

**M. Parthiban and M. Harinath**

**Abstract** In this work, the machining of cemented carbide rods is carried out in wire electric discharge grinding process. It is a hybrid machining process in which machining of micro-parts can be machined with high accuracy, aspect ratios, and flexibility. The rotary axis is mounted on the WEDM machine to produce micro-rods. The workpiece material cemented carbide is selected since it is widely used for cutting tools in industrial applications. This material is tough to be machined through conventional process. From the literature survey, the following three machining parameters such as the input current (A), pulse-off time (μs), and voltage (V) are considered as the input parameters for the machining process. Experimental trials are conducted based on the Taguchi parametric design by machining an Ø2 mm, and length of 10 mm to Ø400 μm with brass wire of Ø 0.25 mm was used to machine the workpiece material under different sets of parameter combinations. The S/N ratio analysis was carried out to find the optimum machining parameters and significant parameters affecting the material removal rate and surface roughness. ANOVA is used to find the most dominating parameter. Finally, it is concluded experimentally and statistically that the significant parameters affecting the material removal rate and surface roughness is input current. By varying these parameters, machining process of cemented carbide is extended to contour profiles.

**Keywords** Wire electric discharge grinding (WEDG) · Design of experiments (DOE) · Analysis of variance (ANOVA) · Cemented carbide (C) · S/N ratio analysis · Material removal rate (MRR)

## 1 Introduction

In recent years, the manufacturing of micro-parts is increased due to the increase in various applications. Hybrid manufacturing machines are used to manufacture more complex and good surface finish parts. The wire electrical discharge machining is

M. Parthiban (✉) · M. Harinath
Department of Mechanical Engineering, PSG College of Technology, Coimbatore, India
e-mail: mpn.mech@psgtech.ac.in



P. Srinivasa Pai and V. Krishnaraj (eds.), *Sustainable Machining Strategies for Better Performance*, Lecture Notes in Mechanical Engineering,
https://doi.org/10.1007/978-981-16-2278-6_14

one such hybrid machining process which uses the transverse wire to manufacture the parts. Because of its high advantages, it is widely used in the aerospace, nuclear, and automotive industries. More complex and intricate shapes can be machined using this machining process. Machining hard materials like titanium with complex shapes can be machined using wire electric discharge grinding process. High precision parts can be fabricated effectively with the help of WEDG process because conventional grinding process takes many hours to grind and polish the materials. The research in this non-traditional machining process was initiated by Kinoshita et al. [1]; they pointed out that the ratio between the clearance and the amplitude of wire vibration is the most appropriate value to judge the short-circuit gap. Masuzawa and Tongshoff [2] developed the WEDG machine and manufactured small pins for micro-parts and microelectrodes and used these microelectrodes as a tool for electrical discharge machining (EDM) applications. Anand [3] in his study, to improve the dimensional accuracy and surface finish of a machining process, carried out optimization with a fractional factorial experiment with an orthogonal array. Masuzawa et al. [4] converted the WEDM into a wire electrical discharge-turning process by placing a submergible water-cooled spindle which can rotate up to 2800 rpm for the production of gear wheels for the gear assembly. Qu et al. [5] studied wire electrical discharge machining for material removal rate by developing a mathematical model for the machining process. Chris Morgan et al. [6] developed the micro-shaft in WEDG process and investigated the straightness error and surface roughness for the machining parameters. Haddad [7] based on the Taguchi design of experiments investigated the effect of machining parameters on surface roughness and roundness in WEDM.

Alireza and Mohammad [8] studied the roundness, surface roughness and MRR in the cylindrical wire electric discharge turning. AISI D3 tool steel is the work material selected, and the experimental design was performed. Parthiban et al. [9] designed and fabricated a rotary axis setup to do an experimental study on wire electrical discharge grinding for difficult to machine materials. Pathiban et al. [10] manufactured microelectrodes of tungsten rod with the help of WEDG process. The process parameters were studied and found that the spindle speed and pulse-off time are the most influencing parameter while manufacturing the microelectrodes. Zhang et al. [11] used microelectro discharge machine to study the micro-dimples and micro-grooves on tungsten carbide-mold substrates with a surface finish of 2 nm. Rao et al. [12] performed parametric analysis in WEDM using Taguchi method on material removal rate and surface roughness for aluminum alloy material. Wang et al. [13] with the advantage of WEDG process, proposed a combined machining process with WEDG technology and electrochemical machining process to fabricate micro pins. With the help of this combined machining processes, micro pins of less than 10 micron were succesffuly fabricated and concluded that the process can be effectively used for producing micro-tools. Pratap et al. [14] fabricated the polycrystalline diamond (PCD) tool in the WEDG process and studied the effects of wire tension and discharge for the manufactured tool. Patnaik et al. [15] used multicut strategy to machine the titanium alloy material in wire electric discharge machining to study the material removal rate, wire wear, and surface integrity. The literature survey indicates the scope for the further research in machining of microcomponents like microprobes, micro-tools in carbide and to manufacture microelectrodes with the use of WEDG process.

The objective of this work is to study the process parameters of the wire electric discharge grinding process during the manufacturing of cemented carbide microelectrodes. The manufactured microelectrodes can be further used as a micro-tool for micro-drilling in electric discharge machining and other micro-applications. In this work, cemented carbide is selected as the work material and the machining parameter such as input current, voltage, and pulse-off time are selected to study the output responses of material removal rate and surface finish for the desired input parameters. ANOVA is performed to find which parameter has significant effect over the output responses.

## 2 Experimental Work

In this work, the manufacturing of carbide microelectrodes and optimization is carried for material removal rate and surface roughness. The rotary axis setup is mounted over the wire electric discharge machine (Mitsubishi (FA10S) CNC WEDM) to conduct all the experiments, and the setup is shown in Fig. 1. Cemented carbide cylindrical rod of an Ø2 mm and length of 10 mm was used as work material, and brass wire of Ø0.25 mm was used as electrode to machine the work material. In this study, micro-machining of carbide electrodes is carried using WEDG process. The workpiece is placed over the collect, and the machining parameters are varied for each experiment. The final diameter of carbide microelectrode is reduced to Ø400 μm from Ø2 mm using the WEDG process. The CAD drawing of the workpiece was

**Fig. 1** Rotary axis setup (WEDG)

**Table 1** Parameter settings for cemented carbide

| Parameter | Level 1 | Level 2 | Level 3 |
|---|---|---|---|
| Input current (A) | 4 | 5 | 6 |
| Pulse-off (μs) | 1 | 2 | 3 |
| Voltage (V) | 8 | 12 | 16 |

used to generate the program by Mastercam software, and the generated program was imported to Mitsubishi WEDM machine for conducting the experimental trials. Based on the input parameter levels, the microelectrodes were manufactured and the output responses of the manufactured microelectrodes are studied.

## 2.1 *Selection of Cutting Parameters*

The selection of variables that affect the performance of the WEDM process is identified based on the literature survey and preliminary experiments performed by the author by considering more number of variables and also through specifically considering the parameters affecting the MRR and surface roughness. The factors selected are input current, pulse-off time, and voltage. The input current decides the concentration of the machining process. Pulse-off time represents the time interval between the discharges which plays an important role. Since the current and voltage are reversely proportional, we have taken the voltage as a parameter. As the rotary axis setup is submerged in fluid, the need for dielectric flushing pressure does not have effect on the machining process, so this parameter was neglected in the present study. The values will be assigned directly from the machine library once the material of the workpiece was specified. The selected parameters with their three levels and their values are given in Table 1.

## 2.2 *Surface Roughness Evaluation for Cemented Carbide*

The surface roughness (Ra) was measured using a non-contact 3D profiler roughness tester Talysurf CCI-Lite. It is used to measure the surface roughness for the carbide rod. The surface roughness Ra value was measured in the carbide rod by focusing the center of the profile. Finally, these are recorded and the setup for measuring the surface roughness is shown in Fig. 2.

**Fig. 2** Setup for surface roughness measurement

## 3 Results and Discussion

The plan of experiments is to conduct nine sets of experiments in which the first column was assigned to the input current and the second column to the voltage and the third column to the pulse-off time. In this present study, interaction of factors is to be considered. The outputs to be studied are surface roughness (μm) and the material removal rate MRR ($mm^3$/min). The experiments are conducted for a set of cutting parameters as listed in L9 orthogonal array. The Designed parameters and its response for Cemented Carbide is shown in Table 2. For each process parameter setup, three trials were performed and the average value of the MRR is recorded.

**Table 2** Designed parameter and response for cemented carbide

| S. No. | Current (A) | Voltage (V) | Time off (μs) | MRR ($mm^3$/min) | Ra (μm) |
|---|---|---|---|---|---|
| 1 | 4 | 8 | 1 | 16.56 | 2.87 |
| 2 | 4 | 12 | 2 | 16.96 | 2.80 |
| 3 | 4 | 16 | 3 | 17.48 | 2.62 |
| 4 | 5 | 8 | 2 | 19.63 | 2.32 |
| 5 | 5 | 12 | 3 | 19.87 | 2.20 |
| 6 | 5 | 16 | 1 | 19.50 | 2.13 |
| 7 | 6 | 8 | 3 | 22.31 | 2.41 |
| 8 | 6 | 12 | 1 | 20.45 | 2.34 |
| 9 | 6 | 16 | 2 | 19.98 | 2.28 |

From the results, to find the optimal process parameters, the results were discussed with the use of various tools like ANOVA and S/N ratio to find out the significant effect of machining parameters on material removal rate and surface roughness.

MRR was calculated according to Eq. (1),

$$\mathrm{MRR} = \frac{\pi (d_0 - d)^2}{4t} \times l \quad (1)$$

where $d_0$ is the initial diameter,

$d$ is the final diameter of the workpiece,
$l$ is the length of the machined part, and
$t$ is the time taken for machining.

The surface roughness parameter $R_a$ was measured using a non-contact 3D profiler roughness tester Talysurf CCI-Lite. The cutoff chosen was 0.8 mm and taken at the mid-region of the specimen.

The experiments were done for different sets of machining parameters, and the images of machined component of cemented carbide electrodes have been shown in Fig. 3, and different contour profiles that are machined are shown in Fig. 4.

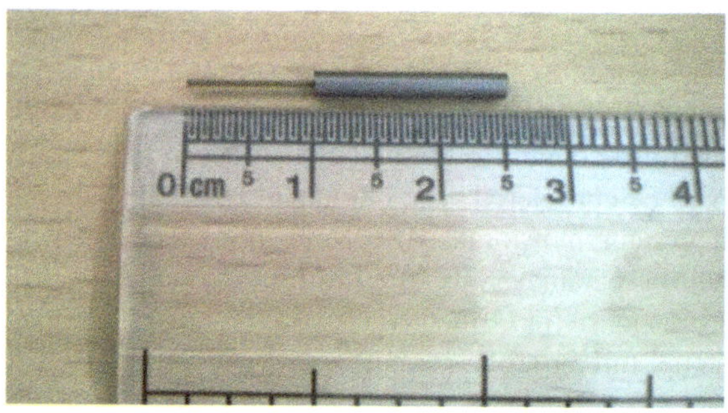


**Fig. 3** Image of machined cemented carbide electrode

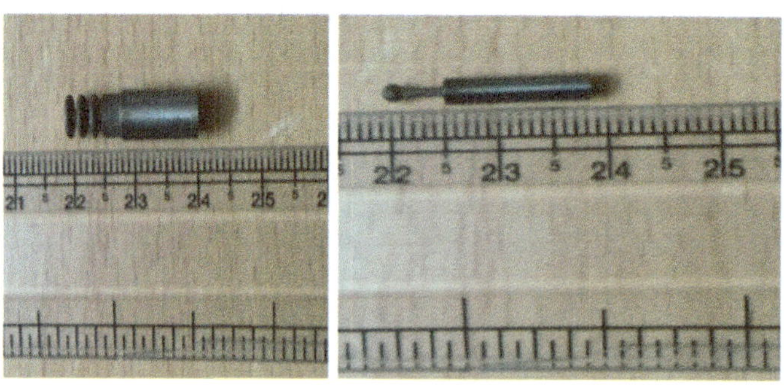


**Fig. 4** Image of contour profile

## 3.1 Data Analysis Using S/N Ratio

From the results, S/N values were calculated for each experiment to find the influence of each factor to the response variables. The results are presented in Figs. 5 and 6. It shows that the greater the difference in S/N values of the control factor, greater

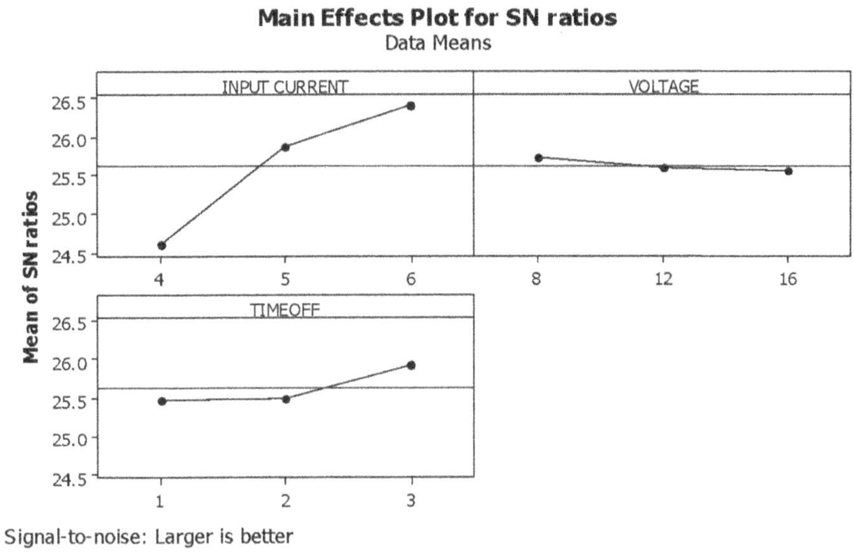


**Fig. 5** Main effect plots of S/N ratio for MRR

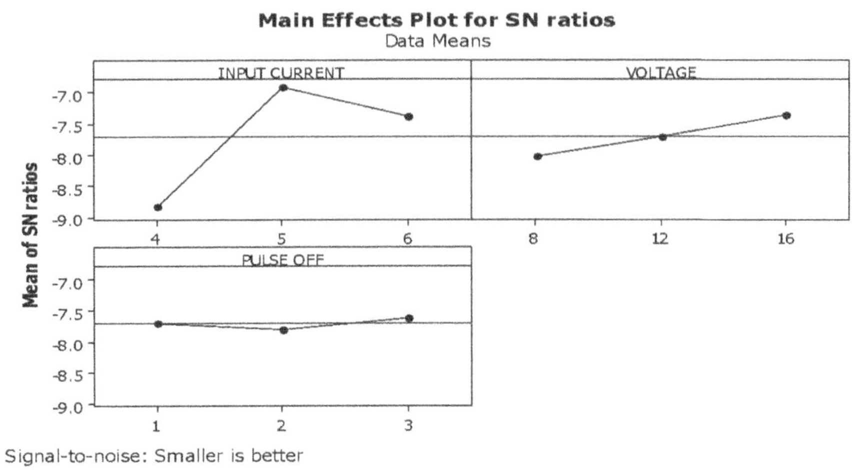


**Fig. 6** Main effect plots of S/N ratio for surface roughness

the significance of the factor to the response variable. Thus, the input current has the strongest influence on MRR and surface roughness.

In order to obtain the optimal machining performance, ‘higher-the-better’ characteristic for material removal rate was taken. Figure 5 shows the highest material removal as attained when input current was 6A; likewise, by analyzing the other two plots of S/N ratio’s, the peak point in the graph provides optimum value of 3(μs) for pulse-off time and 8 V of voltage. attainment of high material removal conditions requires a large discharge of energy by increasing the current and voltage. The diameter of the discharge channel is known to be directly proportional to the pulse-off time. Hence, more craters are produced at short pulse-off time which results in more material removal. Therefore, during the minimum pulse off time the maximum material removal rate is obtained.

Figure 6 shows the variation of S/N ratio for surface roughness for different levels of parameters. The S/N ratio was calculated based on the phenomena of ‘smaller-the-better’ characteristic. The peak values of parameters in the graph have high S/N ratio value which will effect less surface roughness. The factor that has the most influence on surface roughness is again the input current, while the optimal factor values are those with the highest S/N value as current 5 A, voltage 16 V, pulse-off time 3 μs as shown in Fig. 6.

## 3.2 Analysis of Variance (ANOVA)—MRR

**MRR**—Since the ANOVA was done with a confidence level of 95% ($\alpha = 0.05$), the parameter that has a $P$ value less than 0.05 can be taken as the most influencing parameter. From the results given in Table 3, it can be seen that the input current contributes the most for the material removal rate when compared to voltage and pulse-off time. The most influencing parameter is the input current with contribution of 87.16% followed by the voltage of 7.86% and pulse-off time with 1.60%.

**Surface roughness**—Since the ANOVA was done with a confidence level of 95% ($\alpha = 0.05$), the parameter that has a $P$ value less than 0.05 can be taken as the most influencing parameter. From the results given in Table 4, it can be seen that the input current contributes the most for the surface roughness when compared to the other

**Table 3** ANOVA for cemented carbide (MRR)

| Source | DF | Seq SS | Adj SS | Adj MS | $F$ | $P$ | % Contribution |
|---|---|---|---|---|---|---|---|
| Input current (A) | 2 | 23.979 | 23.979 | 11.9897 | 25.83 | 0.037 | 87.16 |
| Pulse-off (μs) | 2 | 0.4403 | 0.4403 | 0.2201 | 0.47 | 0.678 | 1.60 |
| Voltage (V) | 2 | 2.1638 | 2.1638 | 1.0819 | 2.33 | 0.300 | 7.86 |
| Error | 2 | 0.9285 | 0.9285 | 0.4642 | | | 3.37 |
| Total | 8 | 27.512 | | | | | 100 |

**Table 4** ANOVA for surface roughness

| Source | DF | Seq SS | Adj SS | Adj MS | *F* | *P* | % Contribution |
|---|---|---|---|---|---|---|---|
| Input current (A) | 2 | 0.4612 | 0.4912 | 0.54564 | 514.14 | 0.002 | 89.28 |
| Pulse-off (μs) | 2 | 0.0542 | 0.0542 | 0.2714 | 56.81 | 0.017 | 9.82 |
| Voltage (V) | 2 | 0.0049 | 0.0049 | 0.00247 | 5.19 | 0.162 | 0.89 |
| Error | 2 | 0.0009 | 0.0009 | 0.00047 | | | |
| Total | 8 | 0.5514 | 0.5514 | | | | 100 |

parameters. With 89.28% of contribution, the most dominating parameter is the input current followed by pulse-off time with 9.82% and voltage with 0.89%.

## 4 Conclusion

In this study, micro-machining on cemented carbide was carried out for material removal and surface roughness. The experimental investigation has been done on the effects of machining parameters, namely the input current, pulse-off time, and voltage on MRR and surface roughness for micro-machining of carbide rods by WEDG. Experimental trials were taken for full factorial design in order to analyze the selected process parameters for MRR and surface roughness.

- The results show that the manufacturing of carbide microelectrodes of up to Ø 400 μm can be achieved in WEDG process, and by using this process, micro-tools and microelectrodes could be manufactured more effectively.
- By performing ANOVA, the most significant factor is determined, and by analyzing the S/N ratio, the optimal parameter to achieve MRR is when input current at 6A, pulse-off time at 3(μs), and voltage at 8 V. And, the optimal parameter to achieve the surface roughness is when the current is at 5 A, voltage at 16 V, pulse-off time at 3 μs, respectively.
- The most dominating parameter for MRR is input current with a percentage contribution of 87.19%. The second important parameter is voltage with the percentage contribution of 7.86% followed by pulse-off time of 1.60%, respectively.
- The most dominating parameter for surface roughness is input current with a percentage contribution of 89.28%. The second important parameter is pulse-off time with the percentage contribution of 9.82% followed by voltage of 0.89%, respectively.

## References


1. Kinoshita N, Fukui M, Shichida H, Gamo G, Sata T (1976) Study on EDM with wire electrode gap phenomena. Annal CIRP 25(1):141–145
2. Masuzawa T, Fujino M, Kobayashi K (1985) Wire electro-discharge grinding for micromachining. CIRP Ann 34(1):431–434
3. Anand KN (1996) Development of process technology in wire-cut operation for improving machining quality. Total Qual Manag 7(1):11–28
4. Masuzawa T, Okajima K, Taguchi T, Fujino M (2002) EDM-Lathe for micromachining. Annal CIRP 51(1):355–358
5. Jun Q, Shih AJ, Ronald Scattergood O (2002) Development of the cylindrical wire electrical discharge machining process, Part1: Concept, design and material removal rate. Trans ASME 124:702
6. Morgan C, Shreve S, Ryan Vallance R (2003) Precision of micro shafts machined with wire electro-discharge grinding. In: Proceedings of the ASPE winter topical MTG on machines and processes for micro-scale, meso-scale fabrication, metrology and assembly
7. Haddad MJ (2006) Statistical analysis of the results of the cylindrical wire electrical discharge turning process. M.Sc. dissertation, Mechanical Engineering Department, Isfahan University of Technology, Isfahan, Iran
8. Alireza FT, Mohammad JH (2008) Investigation of cylindrical wire electrical discharge turning (CWEDT) of AISI D3 tool steel based on statistical analysis. J Process Technol 198(1–3):77–85
9. Parthiban M, Manigandan C, Muthu Venkadesh G, Ranjith Kumar M (2013) Development of rotary axis for Wire Electrical Discharge Machining (WEDM). Int J Eng Res 2(4):314–317
10. Parthiban M, Krishnaraj V, Naveen Anthuvan R (2014) Optimization of parameters for diameter accuracy in wire electric discharge grinding for micro machining of tungsten rods. Appl Mech Mater 592–594:625–629. ISSN: 1662-7482
11. Zhang Z, Peng H, Yan J (2013) Micro-cutting characteristics of EDM fabricated high-precision polycrystalline diamond tools. Int J Mach Tools Manuf 65:99–106
12. Rao PS, Ramji K, Satyanarayanan B (2014) The experimental investigation and optimization of wire EDM parameters for surface roughness, MRR and white layer in machining of aluminum alloy. Procedia Mater Sci 5:2197–2206
13. Wang JY, Sheu DY (2016) Developing a process chain with WEDG technology and pulse ECM to fabricate ultra micro pins. Procedia CIRP 42:815–818
14. Pratap A, Patra K (2019) Effects of electric discharge dressing parameters on polycrystalline diamond micro-tool surface topography and their micro-grinding performances. Int J Refract Metals Hard Mater 82:297–309
15. Patnaik P, Datta S, Mahapatra SS (2019) WEDM performance of Ti-6Al-4V: emphasis on multi-cut strategy, effects of electrode wire. Mater Today: Proc 18(Part 7):4102–4110

# Exploration of Effectiveness of Ionic Liquid Adopted as an Additive to the Vegetable Oils

**Harpreet Singh, Balraj Singh, and Roshan Lal Virdi**

**Abstract** Abrasive material removal processes are quite challenging due to high temperatures and power requirements. To achieve quality products, unwanted material removal is required from the base processed material. In the process of removing unwanted material from the workpieces, different operations are involved such as threading, grinding, and turning operations. The main parameters of the material removal process are feed, speed, and depth of cut. During the material removal process, heat is generated from the workpiece as well as tool. Therefore, time to time, lubrication is required to tackle with the problems that rise from generation of heat. Minimum quantity lubrication is an alternate of general cooling systems to protect the environment as well as human health. In the present investigation, attempt has been made to explore the surface roughness, grinding forces, specific grinding energy, grinding temperature, and G ratio in MQL grinding of AISI 52100 steel. The three input variables are considered in this investigation, namely vegetable oils, air pressure, and an ionic fluid. Better results have been attained with the MQL grinding by using coconut oil under different parameters. Signal-to-noise (S/N) ratio as per Taguchi design revealed MQL as significant parameters for surface roughness, whereas feed can be set within the range. Taguchi optimized conditions were validated through multiple response optimization using desirability function.

**Keywords** Surface grinding · Vegetable oils · Ionic fluids · Environment-friendly · Inexpensive · Minimum quantity lubrication

## List of Abbreviations

| | |
|---|---|
| MQL | Minimum Quantity Lubrication |
| ILs | Ionic Fluids |
| BMIM $Tf_2N$ | 1-Butyl-3-methyllimidazolium bis(trifluoromethylsulfonyl)imide |
| S/N (or SN) Ratios | Signal to Nose Ratios |

H. Singh (✉) · B. Singh · R. L. Virdi
Department of Mechanical Engineering, Punjabi University, Patiala 147002, Punjab, India



P. Srinivasa Pai and V. Krishnaraj (eds.), *Sustainable Machining Strategies for Better Performance*, Lecture Notes in Mechanical Engineering,
https://doi.org/10.1007/978-981-16-2278-6_15

| | |
|---|---|
| G ratio | Grinding Ratio |
| $V_{wp}$ | Volume of Workpiece Removed |
| $V_{wear}$ | Volume of Wheel Wear |
| $E_t$ or U | Specific Grinding Energy |
| $F_t$ | Tangential Force |
| $F_n$ | Normal Force |
| $v_s$ | Peripheral Speed of Grinding Wheel |
| a | Grinding Depth |
| b | Grinding width |
| Ra | Surface Roughness |

# 1 Introduction

Grinding is a category of machining and utilized as a true metal cutting process. Every fragment of abrasive works as a microscopic single-point cutting edge and has a high negative shape angle. Surface grinding is a class of grinding process employed in various industries to make the workpiece exterior smooth for a better finish [1]. In the grinding operation or material removal process, heat generation is immense in both the job and the grinding wheel which affects the work quality. To minimize these thermic effects, the various kinds of vegetable oils because of their high biological degradability and non-toxic nature could be consumed as lubricants. Vegetable oils (ground nut oil, coconut oil, canola oil) were adopted on the account of their diverse characteristics like tolerable expenditure, low-heat creation within grinding section, and low firmness [2] (Fig. 1).

Ionic fluid is adopted because of its temperature invariability and non-mixing nature in water. It is a salt in the form of liquid at the temperature less than

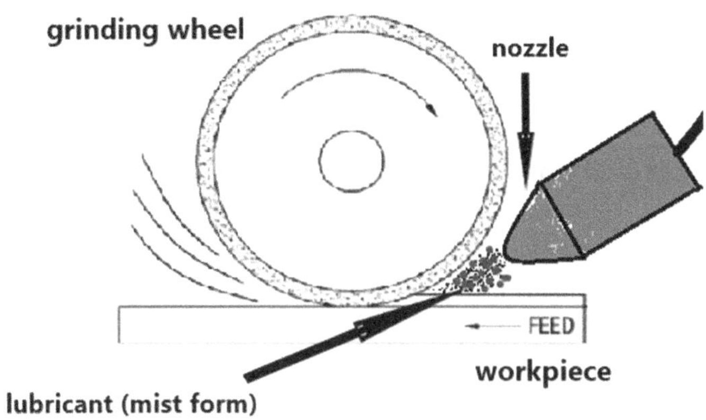


**Fig. 1** Grinding process

100 °C. It is made up of ions comprising cations and anions in bulk and asymmetry [3]. The ionic liquid adopted in this analysis is 1-butyl-3-methylimidazolium bis(trifluoromethylsulfonyl)-imide. If we look at the other gases in ionic fluid gas clarification, *membrane hydrogen is less likely to dissolve than the methane although carbon dioxide* is more able to dissolve than methane and the change in temperature has the principal impact on the breaking up of methane at low temperature. Methane disintegrates less as compared to ethane, butane, and propane [4]. Ionic fluids have the considerable impact in machining operation even where the conditions are tribological. Their application to metalworking fluids in almost dry machining case is greatly probable and can be helpful method to descend forces [5]. The attention has been inclined to the ionic fluids due to their environment-friendly nature as they are reliable, re-utilizable and can go back to the earth without deteriorating it [6]. Khalilpourazari and Khalilpourazary [2] toiled (worked hard) on the optimization technique of the determinants of surface grinding operation and propounded Taguchi method in comparison with other prevailing (or superior) methods rendered the successive consequences. Dambatta et al. [7] in their scrutiny conducted an experimental review with the aid of Taguchi pattern of examination to check the up-shots of granulating determinants and diverse sorts of coolants throughout the course of crushing of Si3N4 improved engineering ceramics and outcomes of the process constants on the grinding forces and exterior grade had been investigated. Besides, from the Taguchi-based analysis the S/N ratio of outcomes were utilized to attain the optimum variable settings.

Goindi et al. [5] in their examination targeted the ionic fluids to be adopted as a lubrication preservative to brew with the vegetable oil (canola oil) in MQL milling to perform on the surface of plain carbon steel and found out that the ionic fluids alter the tribological circumstances of machining approach. Goindi et al. [8] in their exploration put forward the up-shots attained in the comprehensive probe conducted to assess the functioning of oil-free ionic fluid containing fluorine and phosphonium ion-based oil compatible ionic fluids brewed to canola oil or polyethylene glycol in minuscule volumes in obstructed MQL machining of a carbon steel of plain medium under disparate cutting measurands. Ngo et al. [9] perceived in their interpretation that the alkylimidazolium-bis(trifluoromethylsulfonyl)-imide salts portrayed fine thermic invariability when set parallel to alkylimidazolium tetrafluoroborate salts. Lu et al. [10] in their experiment dovetailed (combine) and deduced the ionic liquid named as 1-ethyl-3-hexylimdazolium-bis(trifluoromethylsufonyl)-imide as coolant for the steel-steel contiguity.

Abdul Sani et al. [11] in their review studied the interrelated efficacies of the ionic fluids: the first being based on ammonium and the other ionic fluid having phosphonium as foundation element capable of blending in oil and biologically acceptable adopted as lubrication preservative in antithetical (conflicting) oil at distinctive treat rates in direction of unveiling and perceiving their lubrication potentialities. Ferri et al. [12] in their work aimed to explore the output obtained on the employment of cutting liquids miscible with the ionic fluids. The output to check is of declining the sliding effect of machine equipment and the abrasion of the same. Davis et al. [13] intended to quest the efficacy of ionic fluid (IF), a variety of low melting point salt (below 100 °C), functioning as an accompaniment for executing MQL practice during titanium machining and resulted in curtailing of tool erosion, achieving low cutting forces and acceptable outer finish of work surface. Kalita et al. [14] effectuated (or brought about) an exploration on minimum quantity lubrication grinding to explore the possibility of putting on record the effectiveness of approach of MoS2 nano-lubricants grounded on oils on cast iron and EN steel and considered three dominant variables to deduce the practice effectiveness named as specific energy, rubbing coefficient in grinding, and grinding ratio (G ratio). Li et al. [15] probed the similarities and dissimilarities between the efficacy of MQL grinding by adopting castor oil, rapeseed oil, sunflower oil, palm oil, soybean oil, corn oil, and peanut oil as starting oils on the basis of values of grinding force, grinding temperature, and energy ratio coefficient of MQL grinding attained following the execution of seven unidentical varieties of vegetable oils.

Awale et al. [16] mainly targeted in their study to explore practically the cooling and the lubrication influence of dry, flood, MQL grinding technique on the different measurands named as specific grinding energy, specific grinding force, grinding force ratio, ground surface quality, micro-hardness, and the debris configuration of hardened H13 hot die steel. Irani et al. [17] in his study tell about the systems that have application of cutting liquid high results in the high material removal rates or deepest depth of cut. So, in the long way, the ways adopted in this experimentation are potentially good to make grinding field modern. Barczak and Batako [18] in his probe intended to surge the physical operations understanding performing in the MQL. Shokrani et al. [19] in this exploration took survey on those materials at which machining is not easy, categorized them into three major classes, namely hard materials, ductile materials, and heterogenous materials, and found out those characteristics of those materials which make machining of these materials tough.

From the above discussion, it is very clear that various researchers have done work on vegetable oil with MQL. But, work with vegetable oil with ionic liquid under minimum quantity lubrication (MQL) process is limited. So, significant approach for investigating the performance characteristic of vegetable oils with ionic liquid must be required. This study aims to present the grinding performance of AISI 52100 alloy steel with biodegradable vegetable oil and ionic liquid.

## 2 Experimental Review

The scrutinization was operated on the machine popular by name as surface grinder. The white aluminum oxide grinding wheel (wheel diameter = 180 mm; wheel width = 13 mm; wheel bore diameter = 31.50 mm; wheel grade −80) is equipped in this experiment. The grinding parameters are encapsulated in Table 1.

**Table 1** Grinding parameters

| Description | Characteristics |
|---|---|
| Grinding type | Surface grinding |
| Material adopted | AISI 52100 |
| Grinding environment | MQL, flood, dry |
| Cutting fluids | Groundnut, coconut, canola |
| Lubricant preservative | Ionic fluid |
| MQL flow rate | 100 ml/s |
| Nozzle angle | $15^{o}$ |
| Wheel speed | 28 m/s |
| Depth of cut | 20-μ |
| Grinding wheel passes | 20 |
| Compressed air pressure | 3, 4, 5 (bar) |
| Flood cutting fluid | Water-soluble servo oil |
| Flood cutting flow rate | 8000 ml/h |

## 2.1 Experimental Materials

Groundnut, coconut, and canola oil are three vegetable oils selected in this experiment due to their bio-degradability and non-polluted characteristics. Each oil is employed in three different manners, i.e., pure oil (without mixture), oil + 1% ionic fluid, and oil + 1.5% ionic fluid. The ionic liquid adopted in this analysis is 1-butyl-3-methylimidazolium bis(trifluoromethylsulfonyl)-imide. A high carbon chromium alloy steel (AISI 52100 steel) having density of 7.81 g/cm$^3$ was employed in this analysis. The chemical configuration of AISI 52100 steel is summed up in Table 2.

**Table 2** Chemical configuration of AISI 52100

| Materials | | AISI 52100 |
|---|---|---|
| Ferrous (%) | | Balance |
| Carbon (%) | | 0.98–1.1 |
| Manganese (%) | | 0.25–0.45 |
| Phosphorus (%) | | 0.025 |
| Sulfur (%) | | 0.025 |
| Silicon (%) | | 0.15–0.30 |
| Chromium (%) | | 1.3–1.6 |

**Table 3** Response table for MQL

| Oils | ILs (%) | $P$ (bar) | $F_t$ (N) | $F_n$ (N) | Temp (°C) | Ra (μm) | $U$ (J/mm$^3$) | $G$ ratio |
|---|---|---|---|---|---|---|---|---|
| Groundnut | 0 | 3 | 6.75 | 26.79 | 77.50 | 0.46 | 24.23 | 5.90 |
| Groundnut | 1 | 4 | 6.28 | 25.30 | 70.00 | 0.39 | 22.54 | 6.30 |
| Groundnut | 1.5 | 5 | 5.87 | 24.63 | 68.25 | 0.33 | 21.07 | 6.70 |
| Coconut | 0 | 4 | 6.66 | 23.56 | 70.25 | 0.42 | 23.91 | 6.30 |
| Coconut | 1 | 5 | 6.17 | 22.85 | 67.75 | 0.37 | 22.15 | 6.50 |
| Coconut | 1.5 | 3 | 5.69 | 20.77 | 62.00 | 0.29 | 20.43 | 7.00 |
| Canola | 0 | 5 | 8.96 | 29.98 | 75.00 | 0.61 | 32.16 | 5.60 |
| Canola | 1 | 3 | 8.30 | 28.51 | 74.50 | 0.52 | 29.80 | 5.70 |
| Canola | 1.5 | 4 | 6.87 | 25.05 | 69.25 | 0.37 | 22.80 | 6.25 |

### 2.2 *Experimental Condition*

In this study, three categories of parameters such as vegetable oils, air pressure, and ionic fluid were considered. All these constants were allocated as per the orthogonal ($L_9$) array. For every single observation, the calculation of relative magnitudes of tangential and vertical forces, temperature, surface roughness, grinding energy, and G ratio are done.

## 3 Results and Discussion

### 3.1 *Investigative Strategies and Assessments*

Grinding forces (tangential and normal force) were assessed on the 610B model dynamometer. A total of 9 similar work sections were passed for overall 180 times having each piece passed for 20 times, and eventually, the average of grinding force at each pass had been derived to attain the relative grinding force. The temperature of workpiece surface was checked after every five passes with the aid of a digital thermocouple. At the end, the surface roughness of every single piece was noted by moving tip of Ra tester across the direction of grinding.

**Tangential and Normal Forces**

Tangential force exerts in the direction of tangent to the point where the grinding wheel makes contact with the work section, while grinding and normal force exerts at the right angle to the work section where grinding wheel makes contact with the work section.

### Grinding Temperature

Most of the time, the temperature in the midst of grinding approach is delivered by escalating sliding powers or deleting powers. The coolants are employed on the grounds in many varieties for controlling the temperature as the grinding wheel is furthermore pulverized due to high temperature by elevating tool wear.

### Surface Roughness

Surface roughness most oftenly considered as roughness, is a section of exterior quality. It functions a paramount contribution in commuting the way the practical object will have a relationship with the surroundings. It is unacceptable in nature to achieve a high outer waviness; it is probably a stumbling block and costly to limit it in manufacturing. So, the three variables namely three disparate vegetable oils i.e. groundnut, coconut and canola, air pressure at distinguished values and ionic liquid mixed in vegetable oils are employed in this research to constraint the surface roughness.

### Specific Grinding Energy

The power exhausted in removing unit volume of a material from the work region while grinding is called the grinding energy, which can be measured by applying the following formula:

$$U = \frac{P}{Qw} = \frac{Ft \times Vs}{Vw \times a \times b} \tag{1}$$

where $U$ = specific grinding energy (J/mm$^3$), $Ft$ = tangential force (m/s), $Vs$ = wheel velocity (m/s), $Vw$ = work feed rate (m/s), $a$ = width of working zone (mm), $b$ = depth of cut (mm).

### Grinding Ratio

It is the ratio of volume of material removed ($V_{\text{wp}}$) to the volume of grinding wheel wear ($V_{\text{wear}}$) under the similar grinding conditions. It is illustrated as

$$G\,\text{ratio} = \frac{V_{\text{wp}}}{V_{\text{wear}}} \frac{V_{\text{wp}}}{V_{\text{wear}}} \tag{2}$$

where $V_{\text{wp}}$ = volume of workpiece removed and $V_{\text{wear}}$ = volume of grinding wheel wear.

### 3.2 S/N Ratio

In this optimization analysis, the grade of variables' function could be considered based on three classes, namely, the smaller-the-better, higher-the-better, and nominal-the-better. All the determinants are considered best according to smaller-the-better except for the *G* ratio which is optimum if the S/N ratio analysis provides the value according to the larger-the-better.

No matter what is the class of standard characteristics, the better grade can be accomplished by setting the highest S/N ratio with the operation determinants.

Figures 2, 3, 4, 5, 6, and 7 illustrate the main effects plot for SN ratios of different output determinants. By looking at Figs. 2, 3, 4, 5, 6, and 7, we can conclude that the optimum values for distinct output parameters are obtained for coconut oil, the mixture having ionic fluid 1.5% and at the pressure of 4 bar. As coconut oil has low viscosity which allows this oil to flow smoothly over the work surface due to which the debris from the surface and the wheel pores gets flushed easily. This results in low friction between the wheel and the work region, and due to the better flushing off of debris, the surface appearance is smooth.

The SN ratio for the output parameters ends up at high position when 1.5% ionic fluid, containing coolant, is discharged on the work section. The rationale behind this action is the existence of fluorine (lost free in the course of disintegration of ionic fluids in grinding) that makes the bond with the chips, thereby disconnecting the physical contiguity of the tool and the fragment interface.

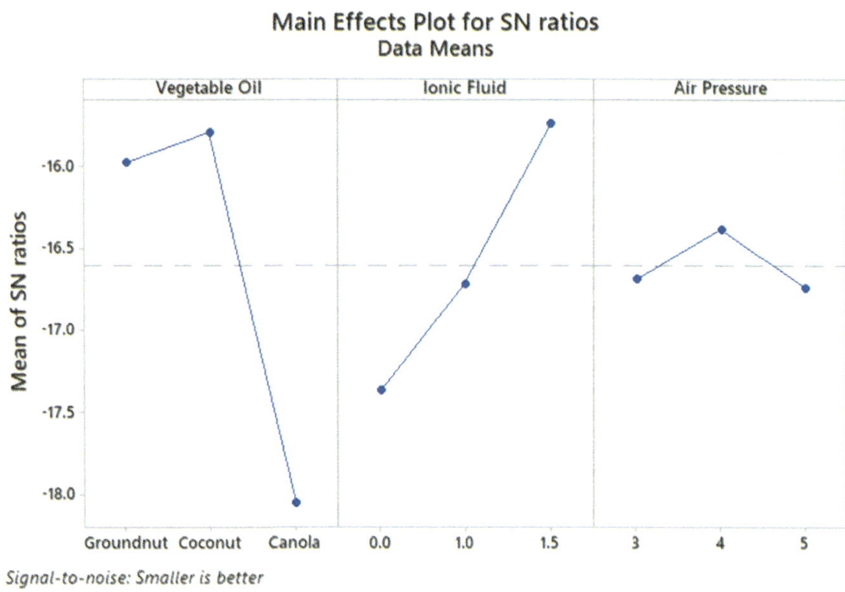


**Fig. 2** Main effects plot for S/N ratios of tangential forces

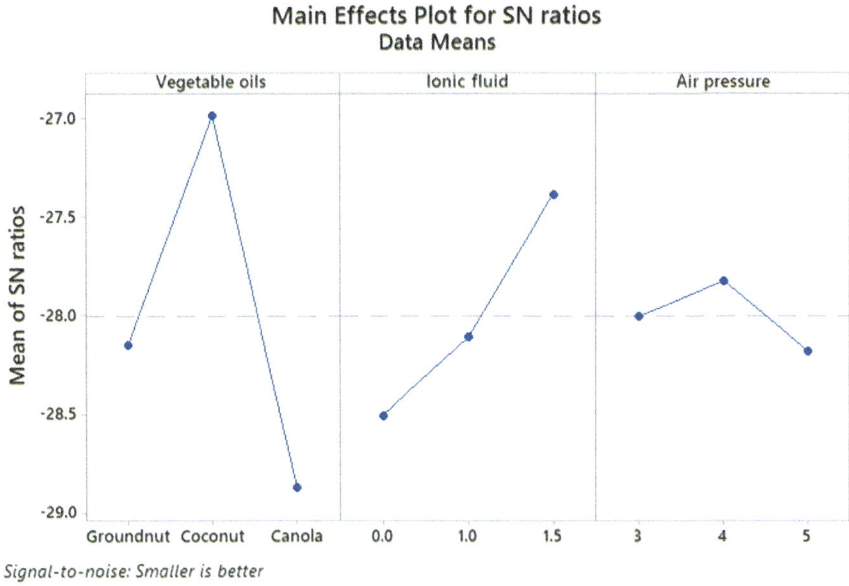


**Fig. 3** Main effects plot for S/N ratios of normal forces

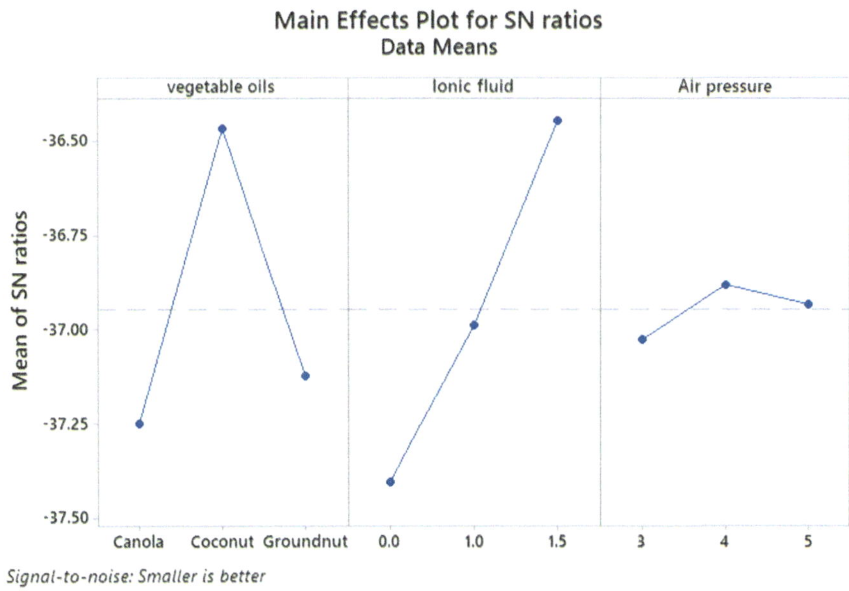


**Fig. 4** Main effects plot for S/N ratios of grinding temperature

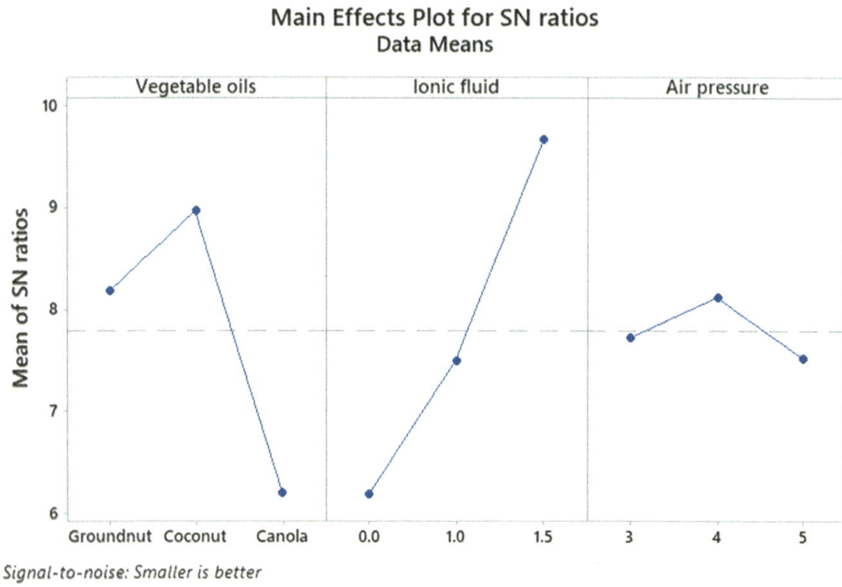


**Fig. 5** Main effects plot for S/N ratios of surface roughness

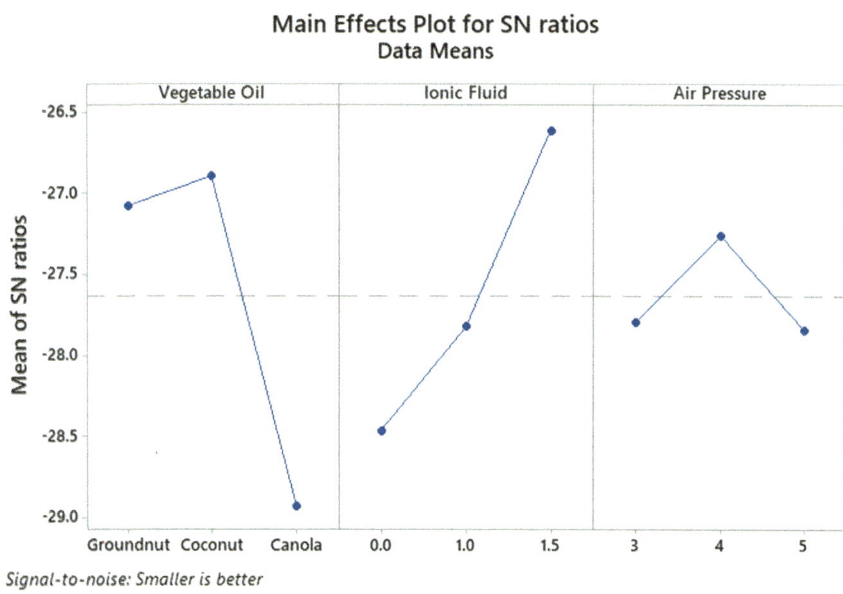


**Fig. 6** Main effects plot for S/N ratios of grinding energy

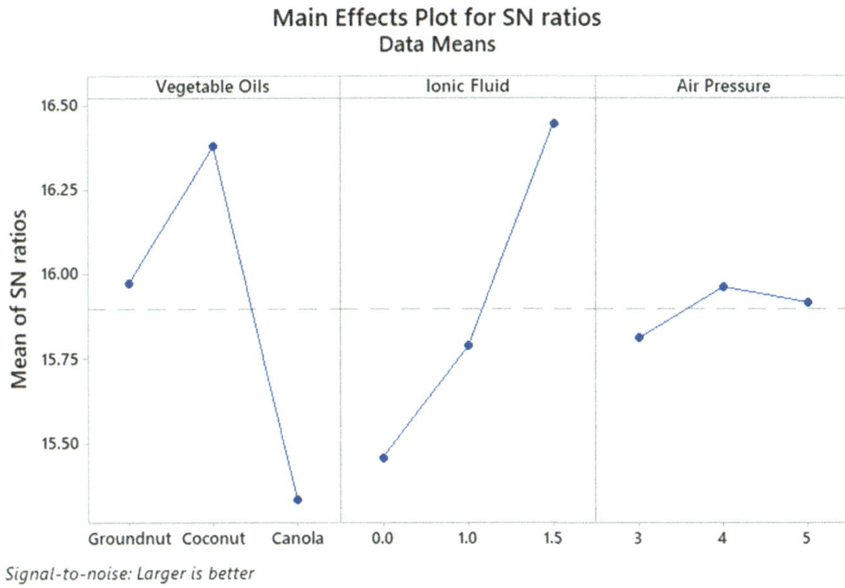


**Fig. 7** Main effects plot for S/N ratios of grinding ratio

In case of air pressure, both the SN ratios for output determinants are finished up at maximum when air is supplied at pressure of 4 bar. It could be an optimum pressure because of the appropriate amount of mist discharged on the work section at this pressure. At 3 bar, the amount of mist discharged to work area could not be satisfactory as the mist discharged may have resulted in less amount that is not enough for building up of satisfactory film for cleaning the debris.

## 4 Conclusions

The entire practice was functioned on AISI 52100 taking into consideration the three amending variables, namely vegetable oils, air pressure, and ionic fluid to deduce the efficacy on the various parameters such as crushing loads, specific granulating energy, grinding temperature, and the exterior unevenness of the job's material.

The epilogues educed by analyzing the entire research are demonstrated below

- Coconut oil is the most optimum oil for lubricating as all the responses are at lower position on usage of coconut oil.
- The proportion of 1.5% ionic fluid in the coolant is the optimal choice for lubrication to get the optimal up-shots. This happens as a result of viscosity change of the coolant after the blending of ionic fluid in lubricant.
- The responses are better at 4 bar pressure when compared with the responses accomplished at 3 and 5 bar pressure.

- The optimal exterior un-flatness accomplished in this analysis is on the employment of coconut oil brewed with the 1.5% ionic fluid as additive of the total amount of coolant at the pressure of 3 bar. It shows that the vegetable oil and ionic fluid are the obligatory factors in this investigation, whereas air pressure is of least importance.
- From the investigation, it can be said that the vegetable oils in MQL technique are authenticated as less deleterious and secure for the surroundings when placed side by side with regular fluids and the expenses are lower than flood grinding as the vegetable oils as coolant are easy to make available at appropriate whereabouts.

To recapitulate, it can be said that the MQL approach is worthy of attention and up to the standard up-shots and can be employed in place of natural machining methods.

## References


1. González-Santander JL, Fernández R, Martín G, Arrazola PJ (2016) A useful analytical formula to avoid thermal damage in the adaptive control of dry surface grinding. Int J Mech Sci 117:152–161
2. Khalilpourazari S, Khalilpourazary S (2018) A robust stochastic fractal search approach for optimization of the surface grinding process. Swarm Evol Comput 38:173–186
3. Zhou F, Liang Y, Liu W (2009) Ionic liquid lubricants: designed chemistry for engineering applications. JCSR 38(9):2590–2599
4. Raeissi S, Peters CJ (2010) High pressure phase behaviour of methane in 1-butyl-3-methylimidazolium bis(trifluoromethylsulfonyl)imide. Fluid Phase Equilib 294(1–2):67–71 (2010)
5. Goindi G, Sarkar P, Jayal A, Chavan S, Mandal D (2018) Investigation of ionic liquids as additives to canola oil in minimum quantity lubrication milling of plain medium carbon steel. Int J Adv Manuf Technol 94(1–4):881–896
6. Vraneš M, Zec N, Tot A, Papović S, Dožić S, Gadžurić S (2014) Density, electrical conductivity, viscosity and excess properties of 1-butyl-3-methylimidazolium bis(trifluoromethylsulfonyl)imide+ propylene carbonate binary mixtures. J Chem Thermodyn 68:98–108
7. Dambatta YS, Sayuti M, Sarhan AAD, Hamdi M (2018) Comparative study on the performance of the MQL nanolubricant and conventional flood lubrication techniques during grinding of Si3N4 ceramic. Int J Adv Manuf Technol 96(9):3959–3976
8. Goindi GS, Jayal AD, Sarkar P (2018) Application of ionic liquids in interrupted minimum quantity lubrication machining of plain medium carbon steel: effects of ionic liquid properties and cutting conditions. J Manuf Process 32:357–371
9. Ngo HL, LeCompte K, Hargens L, McEwen AB (2000) Thermal properties of imidazolium ionic liquids. Thermochim Acta 357–358:97–102
10. Lu Q, Wang H, Ye C, Liu W, Xue Q (2004) Room temperature ionic liquid 1-ethyl-3-hexylimidazolium-bis(trifluoromethylsulfonyl)-imide as lubricant for steel–steel contact. Tribol Int 37(7):547–552
11. Abdul Sani AS, Rahim EA, Sharif S, Sasahara H (2019) Machining performance of vegetable oil with phosphonium- and ammonium-based ionic liquids via MQL technique. J Cleaner Prod 209:947–964
12. Ferri C, Lizarazo S, Troise M, Iglesias P (2018) Ionic liquids as additives to cutting fluids to reduce machine tool friction and wear
13. Davis B, Schueller JK, Huang Y (2015) Study of ionic liquid as effective additive for minimum quantity lubrication during titanium machining. Manuf Lett 5:1–6

14. Kalita P, Malshe A, Kumar S, Yoganath VG, Gurumurthy T (2012) Study of specific energy and friction coefficient in minimum quantity lubrication grinding using oil-based nanolubricants. J Manuf Process 14(2):160–166
15. Li B, Changhe L, Zhang YB, Yaogang W, Dongzhou J, Min Y (2015) Grinding temperature and energy ratio coefficient in MQL grinding of high-temperature nickel-base alloy by using different vegetable oils as base oil. Chin J Aeronaut 29(4):1084–1095
16. Awale A, Srivastava A, Vashista M, Yusufzai MZ (2019) Influence of minimum quantity lubrication on surface integrity of ground hardened H13 hot die steel. Int J Adv Manuf Technol 100(1–4)
17. Irani RA, Bauer RJ, Warkentin A (2005) A review of cutting fluid application in the grinding process. Int J Mach Tools Manuf 45(15):1696–1705
18. Barczak L, Batako A (2012) Application of minimum quantity lubrication in grinding. Mater Manuf Process 27(4):406–411
19. Shokrani A, Dhokia V, Dhokia S (2012) Environmentally conscious machining of difficult-to-machine materials with regard to cutting fluids. Int J Mach Tools Manuf 57:83–101

# Experimental Investigation of Vegetable Oils-Based Minimum Quantity Lubrication Grinding by Using Ionic Liquid

**Balraj Singh, Harpreet Singh, Roshan Lal Virdi, and Khushdeep Goyal**

**Abstract** In the era of modernization, today each and every industry is focusing on improving the quality of the product. Also with the increased requirements of energy utilization and environment protection, minimum quantity lubrication (MQL) has got great attention. As the name suggests, the quantity of the applied lubricant is reduced to its least amount for effective results. Vegetable oils extensively used as base oil in the grinding process due to its non-pollutant and biodegradability properties. This study involves the performances aspects of three vegetable oils (groundnut, rice bran, and palm oil) which were used as cutting fluid to eliminate health hazards related to the operator. Ionic liquid played a significant role in this study, as it exhibits low molten temperature or non-flammable properties. AISI52100 alloys steel used as work material for evaluating the lubrication performance of vegetable oil and ionic liquid. Surface roughness, grinding temperature, grinding forces, grinding ratio, and specific grinding energy have been selected as output parameters. The Taguchi method that uses L9 orthogonal arrays was performed to optimize the process parameter. Results revealed that groundnut oil led to the optimum results as compared to other two oils (Rice bran or palm oil) in terms of forces, temperature and surface roughness. It was concluded that by increasing the concentration of ionic liquid from 1 to 1.5%, there was significance reduction in grinding temperature or grinding forces. Groundnut oil at 4 bar pressure with 1.5% of the concentration of ionic liquid helped to achieve the optimum results.

**Keywords** Environment friendly · Ionic liquid · Minimum quantity lubrication (MQL) · Vegetable oils

## 1 Introduction

Grinding is a machining process which is firstly developed by ancient human by rubbing two rocks together in order to form various weapons and tools. By using

B. Singh (✉) · H. Singh · R. Lal Virdi · K. Goyal
Department of Mechanical Engineering, Punjabi University, Patiala, Punjab 147002, India



P. Srinivasa Pai and V. Krishnaraj (eds.), *Sustainable Machining Strategies for Better Performance*, Lecture Notes in Mechanical Engineering,
https://doi.org/10.1007/978-981-16-2278-6_16

conventional processes, extreme heat is generated at the point of contact of work piece and tool that possess challenge to the researcher to reduce the heat at its acceptable level. Need of environmental protection and human safety has imposed challenge to various research center, universities and industrial sector to find out alternative for conventional machining process [1, 2]. Various cooling and lubrication method has been adopted for minimizing the heat at the grinding zone. However, minimum quantity lubrication technique got great attention due to its number of advantages in terms of safer or cleaner process and ready to use fluid. MQL exhibits eco-friendly and less consumption of power with better lubrication effect. As the name suggest, the quantity of applied lubricant is reduced to its least amount for effective or efficient lubrication. MQL is that techniques in which the usage of lubricant is reduced from liter/minute to milliliter/hours [3]. MQL also known as near dry machining and microlubrication in which small quantity of fluid transported at the interface of tool and work piece for proper functioning of tool [4].

When the cutting tool (grinding wheel) is advance toward the work piece, material is removed in the form of swarf/chips. The chip is produced by the shearing action, and this produced chip may classify into two groups known as primary and secondary shearing zone. In the case of primary shear zone, the material is removed by elasto-plastic deformation, and the most of the energy used in this stage (primary zone) is converted into heat. On the other side, secondary shear zone is responsible for generation of heat or frictional forces by producing the chips/swarf on rake side of the tool. Advancement of the tool toward the work piece may give rise to flank wear on the flank profile of the tool. Machining of advance material is generally associated with lower productivity and higher machining cost. By using hardened material with increased temperature, the chances of shorter tool life or high wear of tool are increased [5–7].

Zhang et al. [8] performed an experiment by using $AL_2O_3$ and sic nanoparticles of different size (30, 50, & 70) with 2% nanofluid by volume fraction. Results revealed that nanofluid (30:70) results in higher material removal rate and also mixture of sic and $AL_2O_3$ increase the efficiency of MQL grinding. Dongzhou, et al. [9] performed an experiment under nanoparticle jet MQL grinding, conventional flood grinding, MQL grinding, and dry grinding condition. Results revealed that nanoparticle jet MQL grinding generates lower specific tangential forces as compared to MQL and dry grinding. Zhang et al. [10] studied the lubricating behavior of mixed nanoparticles and pure nanoparticles. Results indicated that hybrid nanoparticles ($MOS_2$/CNT) obtain better lubrication than pure nanoparticles. Li et al. [11] studied the effect of grinding temperature by using three work material, Ni-based alloys, 45 steel, and nodular cast iron under minimum quantity lubrication cooling (MQLC) for heat transfer. Carbon nanotube (CNT) and nanofluid were used with 2 and 2.5% volume fraction and results indicated that grinding temperature of 2.5% nanofluid was lower than that of 2% nanofluid.

Hadad et al. [12] performed an experiment on 100cr6 steel by using CBN and $AL_2O_3$ grinding wheel. Results showed that MQL process can increase the grind-ability of 100cr6 hardened steel. Guo et al. [13] use castor oil that act as base oil and mixed separately with six different vegetable oil (palm, maize, sunflower, soyabean, peanut, and rapeseed) at 1:1 ratio to alter their rheologic properties using nickel-based GH4169 alloys. Results revealed that viscosity of mixed oil was lower than pure one, and lubricating properties of mixed oil were higher to pure castor oil. Li et al. [14] demonstrated the comparative study of MQL by using vegetable oils on the basis of energy ratio coefficient, grinding forces, and grinding temperature. Results revealed that the castor oil produced lower value of grinding forces among vegetable oils while generating high temperature. Emami et al. [15] they use four different kind of lubricants, viz. vegetable oils, synthetic, minerals, and hydrocracked under MQL grinding. Results revealed that hydrocracked-based oil and synthetic oil gave satisfactory performance under MQL grinding process of $AL_2O_3$ ceramics. Dambatta et al. [16] reported the performance characteristics of flood cooling and MQL during the grinding of $Si_3N_4$ ceramics. Results showed that tangential and normal forces were lower in the case of MQL as compared to flood cooling.

Rabiei et al. [17] studied the mechanical characteristics of work material of two hard steel (100cr6 & HSS) and two soft steel (CK45 & S305) under MQL grinding. Results revealed that MQL techniques decrease the friction coefficient and grinding forces in both hard and soft steels. Moreover, while applying MQL techniques, better surface quality and finish can be obtained. Goindi et al. [18] use 1 wt% of imidazolium-based ionic liquid in canola oil on plain carbon steel to find out the effectiveness and feasibility of ionic liquid under MQL milling. Results revealed that ionic liquid significantly effects the tribology properties and should be used as additives in order to reduce the energy usage or machining force.

Goindi et al. [19] studied the behavior of phosphonium-based oil miscible ionic liquid (IL) and fluorine containing non-miscible ionic liquid which were mixed to polyethylene glycol and canola oil under MQL machining. Results indicated that fluorine consisting IL showed superior performance at higher cutting speed and can be used for machining of high-speed applications. Virdi et al. [20] perform an experiment by using inconel-718 alloy to find out the tribological performance under nanofluid minimum quantity lubrication (NFMQL). Results indicated that nanofluid minimum quantity lubrication significantly enhances the performance in terms of grinding ratio (G-ratio) and surface roughness.

It is clear from the above discussion that researchers have done experimentations in the field of vegetable oil-based nanofluid MQL. But the work done in investigating the performance of ionic liquid MQL technique for grinding of Inconels using vegetable oil-based ionic liquid MQL system is very limited. There is significant scope and need to examine the grinding of hard to machine materials like AISI52100 alloys using vegetable oils. Therefore, the present work aims to examine the behavior of vegetable oil-based ionic fluids in grinding of hard to machine materials using biodegradable vegetable oils.

## 2 Experimental Setup

This experiment was performed by using a surface grinder. Three vegetable oils (Groundnut, Rice bran, and palm oil) were used as lubricant. The various parameters of surface grinder were as follows: Principle axis power 0.75 kw/h, highest rotating speed 2886 r.p.m, table speed 0.66 m/s, and frequency 50 Hz. A clip-type thermocouple device was used for measuring the temperature. A dynamometer was used for measuring the tangential and normal forces (Fig. 1).

### 2.1 Work Piece and Wheel Material

The experiment was performed by using a white $Al_2O_3$ grinding wheel on precision surface grinder. AISI52100 alloys steel was used as base material to perform all the experiments, and the constitution of material is displayed in Table 1. Groundnut

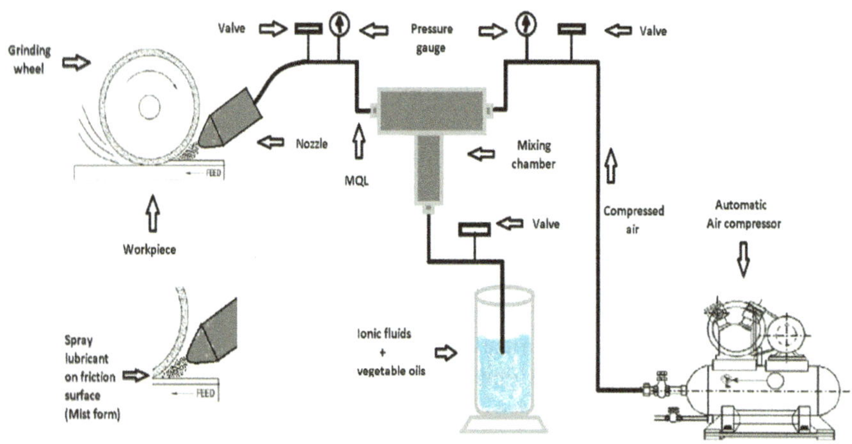


**Fig. 1** Schematic line diagram of minimum quantity lubrication process (MQL)

**Table 1** Composition of material

| Elements | Composition (%age) |
|---|---|
| Cr | 1.14 |
| Fe | 97.13 |
| C | 0.96 |
| S | 0.06 |
| P | 0.047 |
| Si | 0.19 |
| Mn | 0.47 |

oil, rice bran oil, and palm oil are used for conducting all the experiments in the composition of ionic liquid (BMIS TFSI).

## 2.2 Experimental Condition

While performing the experiments, the speed of grinding wheel kept constant which is 28 m/s which means 2886 r.p.m. The flow of air pressure kept 3 bar, 4 bar, and 5 bar while the depth of cut taken as 20 μm by which the tool removes the material from base material in the form of chips. The other grinding parameters shown in Table 2 give necessary data about the process. While performing all experiment, the nozzle distance kept constant. Major three variables used in this research work are:

- Air pressure

**Table 2** Grinding parameters

| Description | Properties |
|---|---|
| Grinding type | Surface grinding |
| Material adopted | AISI 52100 |
| Cutting fluids | Groundnut, rice bran, palm oil |
| Lubricant preservative | Ionic fluid |
| MQL flow rate | 100 ml/s |
| Nozzle angle | 15° |
| Wheel speed | 28 m/s |
| Depth of cut | 20-μm |
| Grinding wheel passes | 20 |
| Compressed air pressure | 3, 4, 5 (bar) |

**Table 3** Response table

| Oils | P, (Bar) | ILs (%) | $F_t$, (N) | $F_n$, (N) | Temp (°C) | Ra (μm) | U (J/mm$^3$) | G-ratio |
|---|---|---|---|---|---|---|---|---|
| Palm | 3 | 1.5 | 7.62 | 27.16 | 74.23 | 0.36 | 29.23 | 6.61 |
| Palm | 4 | 0 | 8.55 | 30.23 | 82.15 | 0.40 | 31.75 | 5.2 |
| Palm | 5 | 1.0 | 8.18 | 29.07 | 79.87 | 0.37 | 31.03 | 4.92 |
| Groundnut | 3 | 0 | 6.61 | 29.12 | 85.13 | 0.31 | 23.75 | 7.63 |
| Groundnut | 4 | 1.0 | 5.76 | 26.13 | 80.27 | 0.27 | 21.41 | 8.69 |
| Groundnut | 5 | 1.5 | 5.28 | 24.45 | 68.21 | 0.25 | 20.17 | 7.69 |
| Rice bran | 3 | 1.0 | 7.23 | 28.10 | 80.97 | 0.33 | 26.13 | 6.81 |
| Rice bran | 4 | 1.5 | 6.69 | 26.09 | 70.12 | 0.28 | 22.06 | 8.44 |
| Rice bran | 5 | 0 | 7.78 | 29.23 | 84.94 | 0.36 | 27.33 | 7.63 |

- Concentration of ionic liquid
- Composition of the vegetable oils.

# 3 Result and Discussion

## 3.1 Analytical Method and Measurement

Grinding forces (Tangential and Normal) were measured by using dynamometer. After the completion of grinding, the surface roughness was calculated by surface roughness tester. The tip of the surface roughness tester was move across the direction of grinding to find out the accurate value of Ra (Table 3).

**Tangential and Normal Forces**
The abrasive wear of work material results is production of various kinds of forces which are mainly characterized as tangential and normal forces. These forces are differentiating on the basis of direction of applied load. Normal forces are those forces which are applied normal to the axis of movement of grinding wheel. Tangential forces are those forces which act along the direction of grinding wheel.

**Grinding Temperature**
Grinding temperature is an essential parameter during the process of grinding as it affects the surface integrity of work material. As the grinding temperature increased, the chances of surface roughness or thermal damage to work piece are also increased. Most of the grinding temperature is increased by friction or cutting forces, so main purpose of this study is to reduce these forces at grinding zone.

**Surface Roughness**
The surface roughness of the ground samples was examined using Mitutoyo SJ-201 roughness tester. The roughness was measured at the center of the work piece by sampling five points with a cut-off length of 2.5 mm. During the process of grinding, the optimum value of surface roughness should lie between 0.1 and 2 μm.

**Surface Grinding Energy (S.G.E.)**
S.G.E is that energy which is required for removing unit volume of material. Specific grinding energy is directly proportional to the product of tangential force and wheel velocity. It is measured by using following equation.

$$\text{Specific Grinding Energy} = \frac{F_t \times V_s}{V_w \times a \times b} \tag{1}$$

where $F_t$ = tangential force (N), $V_s$ = wheel velocity (m/s), $V_w$ = work feed rate (m/s), $A$ = width of working zone (mm), and $B$ = depth of cut (mm).

**Grinding Ratio (G-Ratio)**
Grinding ratio can be defined as the ratio of volume of material removed per unit volume of wheel wear. High grinding ratio is always desirable because it would result in lower wheel wear which results in longer tool life.

$$\text{G-Ratio} = \text{vol. of material removed.}/\text{vol. of grinding wheel wear} \quad (2)$$

## 3.2 *Signal to Noise Ratio Analysis*

While performing the process of optimization and validation, the performance features of process parameter could be categorized into nominal-the-better, smaller-the-better, and larger-the-better. This experimentation was performed to acquire the optimal value, i.e., small value of surface roughness, specific grinding energy, grinding temperature, grinding forces (Ft & Fn), and higher value of grinding ratio through grinding performance process parameters (Figs. 2, 3, 4, 5, 6 and 7).

The graphs depict the mean effect plot for S/N ratio for the tangential force, normal force, surface roughness, grinding temperature, specific grinding energy, and grinding ratio. According to the graph of main effect plot for S/N ratio analysis, the optimum parameter is groundnut oil at 4 bar pressure by using 1.5 concentration

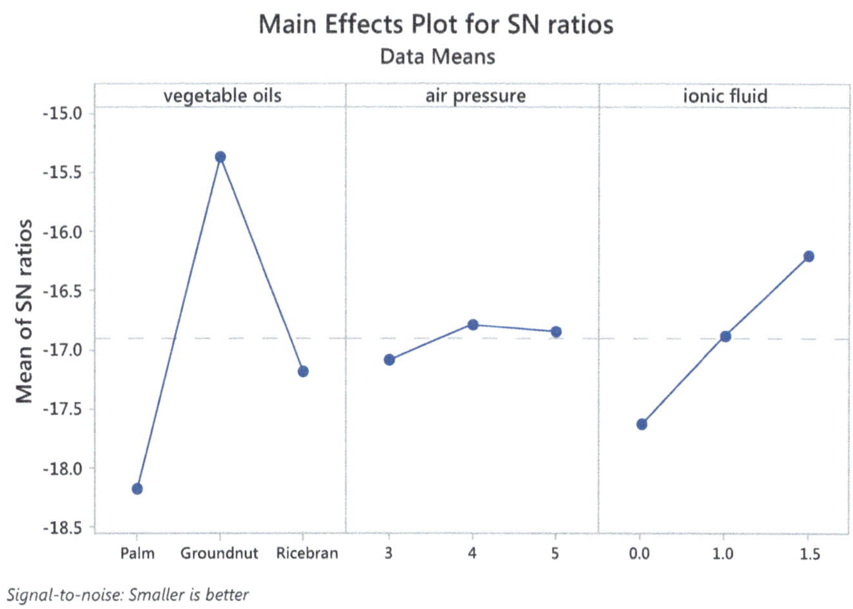


**Fig. 2** Main effect plot for data mean of tangential forces

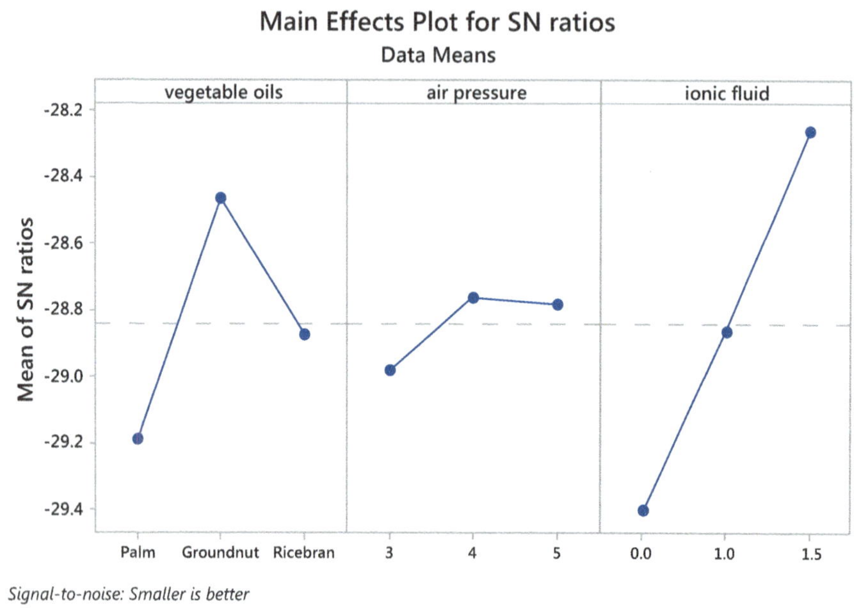


**Fig. 3** Main effect plot for data mean of normal forces

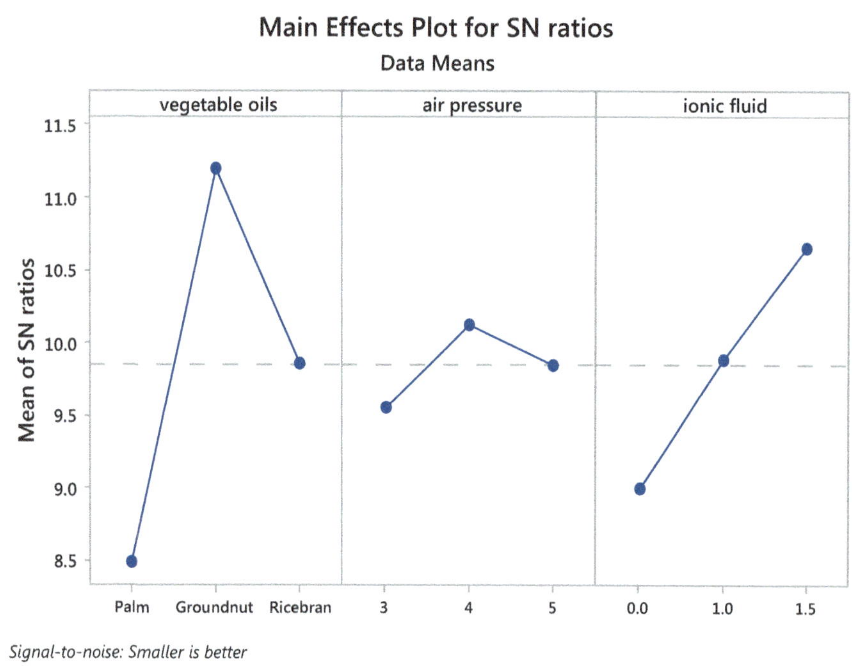


**Fig. 4** Main effect plot for data mean of surface roughness (Ra)

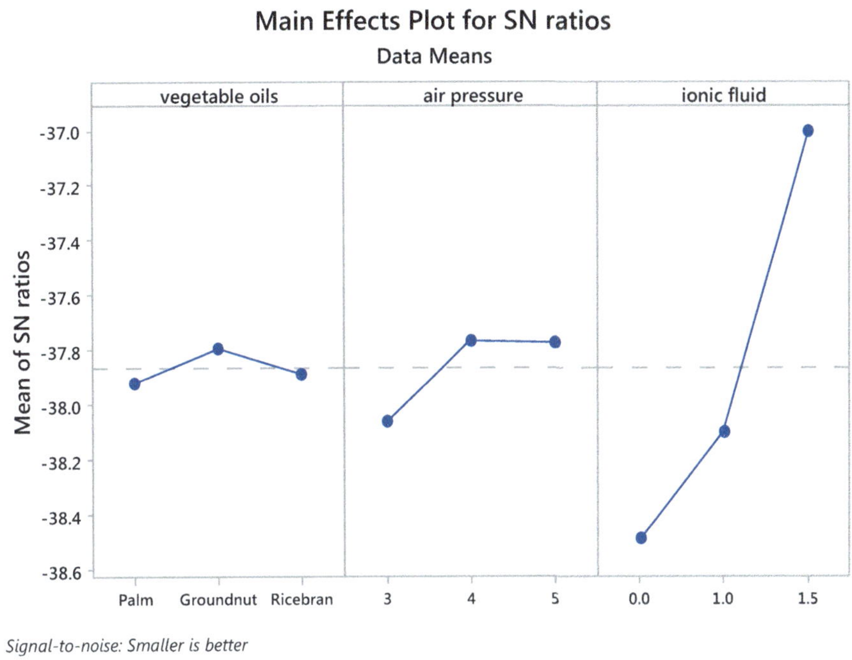


**Fig. 5** Main effect plot for data mean of grinding temperature

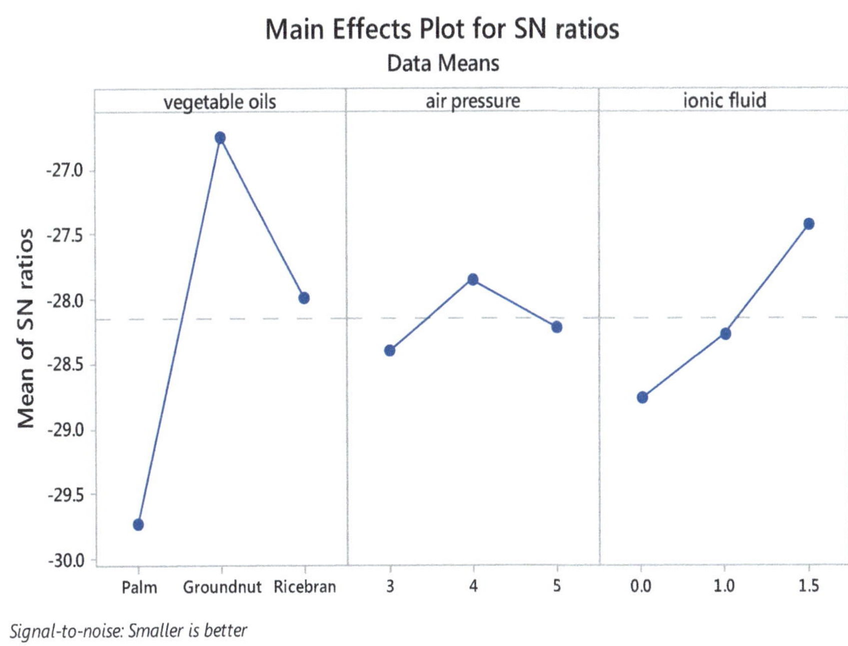


**Fig. 6** Main effect plot for data mean of S.G.E. (U)

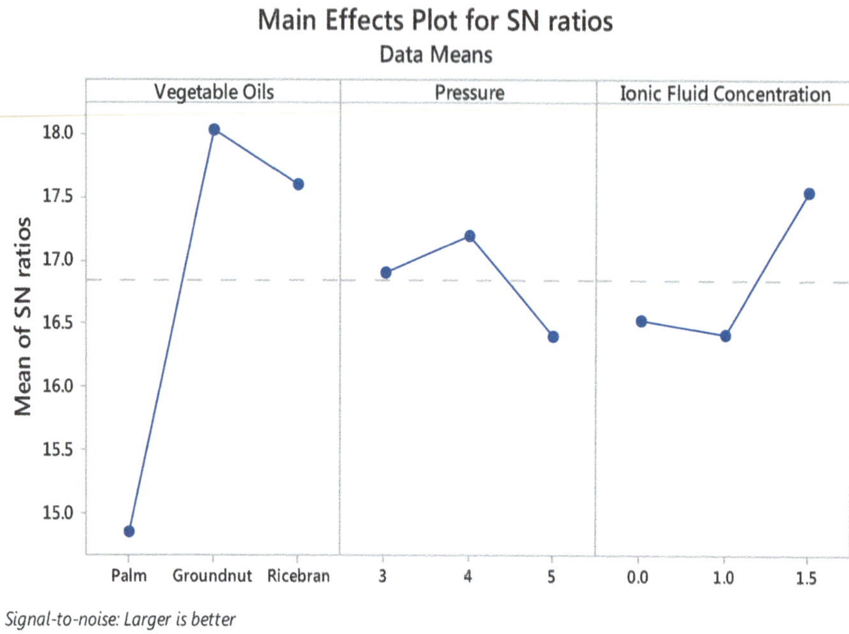


**Fig. 7** Main effect plot for data mean of grinding ratio

of ionic liquid. Study shows that oil with low viscosity results in high fluidity and converse is equally true. Among all three oil (groundnut, rice bran, palm), groundnut oil exhibits low viscosity and oil with low viscosity possess high fluidity. By using groundnut oil, there is fewer clusters of chips on the machined surface of work piece. Also, there is less consumption of energy by using groundnut oil which resulted in less grinding forces (Ft & Fn), temperatures, surface roughness (Ra), specific grinding energy (S.G.E), and high grinding ratio (G-Ratio).

From the graph, it is also observed that mean S/N ratio is low at 3 bar pressure and maximum at 4 bar pressure. This is due to high pressure of air that increased the speed of falling the aerosol droplet (mixture of air and fluid) on the work piece. Also by using the 4 bar pressure, it will generate the stable air oil film at grinding zone which results in efficient lubrication with optimum results. Moreover, graph also depicted that while increasing the pressure from 4 to 5 bar mean, S/N ratio is slightly decreased because the falling of the droplets very rapid, and it is unable to create a stable air fluid layer between the interface of grinding wheel or work piece.

From the main effect plot for data mean graphs, it was also observed that ionic liquid played significant role in order to minimizing the specific grinding energy, grinding forces, surface roughness, and grinding temperature. This is due to the presence of the fluorine in ionic liquid which makes the bond with newly fresh chips/swarf, thereby decreasing the adherence of the chip to the rake profile of the tool which results in lower machining forces. While performing the experiments, higher grinding ratio (G-Ratio) is always desirable because high G-ratio would result into lesser wheel wear or higher tool life.

## 4 Conclusions

Three vegetable oils were mixed with ionic liquid according to orthogonal L9 arrays. Vegetable oils have been used due to its non-pollutant and biodegradability properties. The following conclusions have been obtained from this experiment:

- It is observed that groundnut oil led to the optimum results as compared to other two oils (palm or ricebran) in matter of the forces, G-Ratio, or surface roughness (Ra). The results indicated that pressure 4 bar was the optimum level for optimum result because high pressure of air increases the speed of falling of aerosol at grinding zone which reduces the grinding forces.
- The concentration of ionic liquid has also affected the results. It was observed that by increasing the concentration of ionic liquid from 1 to 1.5%, there was significance reduction in specific grinding energy, grinding temperature surface roughness, or grinding forces. Groundnut oil at 4 bar pressure with 1.5% concentration of ionic liquid helped to achieving the optimum results.
- It is, therefore, concluded that minimum quantity lubrication process is the best method in terms of its non-pollutant properties or biodegradability. Therefore, MQL is an attractive replacement of the conventional machining process.

## References

1. Irani RA, Bauer RJ, Warkentin A (2005) A review of cutting fluid application in the grinding process. Int J Mach Tools Manuf 45(15):1696–1705
2. Silva LR, Catai RE, Fusse RY, Franca T, Aguiar P (2005) Study on the behavior of the minimum quantity lubricant–MQL technique under different lubricating and cooling conditions when grinding ABNT 4340 steel. J Brazil Soc Mech Sci Eng 27
3. Barczak LM, Batako ADL (2012) Application of minimum quantity lubrication in grinding. Mater Manuf Processes 27:406–411
4. Boubekri N, Shaikh VA (2012) Machining using minimum quantity lubrication: a technology for sustainability 2(1)
5. Shokrani A, Dhokia V, Newman ST (2012) Environmentally conscious machining of difficult-to-machine materials with regard to cutting fluids. Int J Mach Tools Manuf 57:83–101
6. Childs T, Maekawa K, Obikawa T, Yamane Y (2000) Metal machining: theory and applications. Butterworth-Heinemann
7. Abukhshim N, Mativenga P, Sheikh MA (2006) Heat generation and temperature prediction in metal cutting: a review and implications for high speed machining. Int J Mach Tools Manuf 46(7–8):782–800
8. Zhang X, Li C, Zhang Y, Wang Y, Li B, Yang M, Guo S, Liu G, Zhang N (2016) Lubricating property of MQL grinding of $Al_2O_3$/SiC mixed nanofluid with different particle sizes and microtopography analysis by cross-correlation. Precis Eng 47

9. Dongzhou J, Li C, Zhang D, Zhang YB, Zhang X (2014) Experimental verification of nanoparticle jet minimum quantity lubrication effectiveness in grinding. J Nanoparticle Res 16
10. Zhang YB, Li C, Dongzhou J, Zhang D, Zhang X (2015) Experimental evaluation of the lubrication performance of $MoS_2$/CNT nanofluid for minimal quantity lubrication in Ni-based alloy grinding. Int J Mach Tools Manuf 99:19–33
11. Li B, Li C, Zhang YB, Wang Y, Min Y, Dongzhou J, Zhang N, Wu Q, Ding W (2017) Numerical and experimental research on the grinding temperature of minimum quantity lubrication cooling of different workpiece materials using vegetable oil-based nanofluids. Int J Adv Manuf Technol 93(12–13):1971–1988
12. Hadad M, Tawakoli T, Sadeghi MH, Sadeghi B (2012) Temperatureemperature and energy partition in minimum quantity lubrication-MQL grinding process. Int J Mach Tools Manuf 54–55:10–17
13. Guo S, Li C, Zhang YB, Wang Y, Li B, Min Y, Zhang X, Liu G (2016) Experimental evaluation of the lubrication performance of mixtures of castor oil with other vegetable oils in MQL grinding of nickel-based alloy. J Cleaner Prod 140
14. Li B, Li C, Zhang YB, Yaogang W, Dongzhou J, Min Y (2015) Grinding temperature and energy ratio coefficient in MQL grinding of high-temperature nickel-base alloy by using different vegetable oils as base oil. Chin J Aeronaut 29(4)
15. Emami M, Sadeghi MH, Sarhan AAD, Hasani F (2014) Investigating the minimum quantity lubrication in grinding of $Al_2O_3$ engineering ceramic. J Cleaner Prod 66:632–643
16. Dambatta Y, Sayuti M, Sarhan AAD, Hamdi M (2018) Comparative study on the performance of the MQL nanolubricant and conventional flood lubrication techniques during grinding of $Si_3N_4$ ceramic. Int J Adv Manuf Technol 96(9):3959–3976
17. Rabiei F, Rahimi AR, Hadad M, Ashrafijou M (2015) Performance improvement of minimum quantity lubrication (MQL) technique in surface grinding by modeling and optimization. J Cleaner Prod 86:447–460
18. Goindi G, Sarkar P, Jayal AD, Chavan S, Mandal D (2018) Investigation of ionic liquids as additives to canola oil in minimum quantity lubrication milling of plain medium carbon steel. Int J Adv Manuf Technol 94(1–4)
19. Goindi G, Jayal AD, Sarkar P (2018) Application of ionic liquids in interrupted minimum quantity lubrication machining of plain medium carbon steel: effects of ionic liquid properties and cutting conditions. J Manuf Process 32:357–371
20. Virdi RL, Chatha SS, Sidhu HS (2021) Experimental investigation on the tribological and lubrication behaviour of minimum quantity lubrication technique in grinding of inconel 718 alloy. J Tribol Int 153:106581

# Comparison of Copper and Tungsten Electrodes for the Electric Discharge Machined SUS-316L

**Gurpreet Singh, Amit Mahajan, Sandeep Devgan, and Sarabjeet Singh Sidhu**

**Abstract** The present work reported the performance assessment of two electrodes, namely copper and tungsten for the electric discharge machining of SUS-316L. The experimental work was performed according to Taguchi's methodology of orthogonal array considering electrode, current, pulse-on time, pulse-off time as input parameters. Three output responses, i.e., material removal rate, surface roughness, and microhardness were opted for deciding the significance of input parameters on the machined surface. Furthermore, each response was statistically validated using analysis of variance for investing the dominating factors. It was revealed that the current and electrode were the most significant factors affecting all the three responses. For material removal rate, current 28 A (contribution: 55.58%) and copper as electrode material (contribution: 33.92%) noted as significant factors. The roughness of the electric discharge machined surface directly relates to the intensity of spark generated within the working area. Similar findings were observed in the study, current (contribution: 43.29%), pulse-on time (contribution: 19.06%), and electrode (contribution: 13.60%) were the factors which majorly contribute to the roughness of the machined surface. However, tungsten electrode noted as prominent affecting the microhardness of the machined SUS-316L surface. The sample machined at 28 A of current, pulse-on time 90 μs, and pulse-off time 60 μs exhibits the utmost microhardness value with a maximum contribution by current (contribution: 47.84%), followed by electrode type (contribution: 28.92%).

**Keywords** SUS-316L · Electric discharge machine · Material removal rate · Surface roughness · Microhardness

G. Singh · S. S. Sidhu
Mechanical Engineering Department, Beant College of Engineering & Technology, Gurdaspur, Punjab 143521, India

A. Mahajan (✉) · S. Devgan
Mechanical Engineering Department, Khalsa College of Engineering & Technology, Amritsar, Punjab 143001, India



P. Srinivasa Pai and V. Krishnaraj (eds.), *Sustainable Machining Strategies for Better Performance*, Lecture Notes in Mechanical Engineering,
https://doi.org/10.1007/978-981-16-2278-6_17

# 1 Introduction

Today, non-conventional machining techniques establish their existence in the manufacturing field by machining the hard and complex-shaped materials and alloys with precisely and improved surface quality [1, 2]. The advanced machining methods such as electric discharge machining, laser beam machining, electrochemical machining, and ultrasonic machining have a wide range of applications in the modern industries [3]. Electric discharge machining (EDM) is an enormously renowned fabricating technique amid all other developed machining processes. This machining technique has also an immense impact on the surface properties and develops thick subsurface films with modified elemental composition and microstructure [4]. EDM is also known as an electro-spark machining or thermo-electric technique where the tool electrode derives controlled high-frequency electric sparks that are directed on the workpiece substrate. The millions of electric discharges generate high temperatures on the targeted region which results in the removal of a certain volume of metal from the substrate [5, 6]. EDM process parameters such as pulse duration, current, dielectric medium, voltage, type of electrode, and polarity (negative or positive) play a momentous role in the machining of different materials and alloys.

Guu et al. [7] considered this thermo-electric technique for surface tailoring of AISI D2 tool steel. They reported that by inflating the spark plasma energy within the workpiece and tool electrode, the recast layer's hardness and the thickness could be enhanced. Das et al. [8] machined EN31 tool steel and observed current as a significant factor for material removal rate (MRR) and surface roughness (SR). The amplification in the intensity of current coupled with pulse-on time predominantly affects the MRR. A similar result was reported by Sharif et al. [9] unfolding current as an eminent parameter influencing all the output responses of the EDMed 316L workpiece. The stainless steel 316L (SUS-16L) is majorly employed for the marine industry, heat exchangers, and biomedical applications because of its corrosion resistance behavior in acetic and alkaline chloride environments. The present work devoted to the machining of SUS-316L using electric discharge machining. The machining performance of two electrodes, i.e., copper and tungsten is compared, and their results are assessed in terms of material removal rate, surface roughness, and microhardness.

# 2 Materials and Methods

The workpiece material, i.e., SUS-316L was procured from Metline Industries, Mumbai, India, and used in the form of a rectangular block (size: 75 mm × 40 mm × 5 mm) for conducting the experiments. The properties of the workpiece and electrode materials are listed in Tables 1and 2, respectively.

The experimental design was generated according to Taguchi's $L_{18}$ (2^1 × 3^3) orthogonal array using Minitab-17 software. Table 3 demonstrates the input

**Table 1** Properties of workpiece material

| Properties | Units | Details |
|---|---|---|
| Material | – | SUS-316L steel |
| Chemical composition | – | Cr:16.13%, Ni:10.15%, Mo:2.05%, Mn:1.07%, Si:0.41%, P:0.22%, C:0.016%, and Fe:Remainder |
| Workpiece size | mm | 50 × 100 × 5 |
| Density | g/cm$^3$ | 7.99 |
| Melting range | °C | 1371–1399 |
| Thermal conductivity | W/mK | 16.2 |
| Specific heat capacity | J/g°C | 0.5 |

**Table 2** Properties of electrode materials

| Properties | Units | Details | |
|---|---|---|---|
| Material | – | Electrolytic copper | Tungsten |
| Purity | – | 99.9% | W-98.3%, $ThO_2$-1.7% |
| Diameter | mm | 8.5 | 10 |
| Density | g/cm$^3$ | 8.904 | 19.25 |
| Melting point | °C | 1083 | 3410 |
| Boiling temperature | °C | 2562 | 5530 |
| Thermal conductivity | W/mK | 388 | 173 |

**Table 3** Experimental parameters and their levels

| Process parameter | Units | Symbol | Level 1 | Level 2 | Level 3 |
|---|---|---|---|---|---|
| Electrode | – | A | Tungsten | Copper | – |
| Discharge current | Ampere | B | 20 | 24 | 28 |
| Pulse-on time | μ-seconds | C | 60 | 90 | 120 |
| Pulse-off time | μ-seconds | D | 60 | 90 | 120 |

machining factors, i.e., electrode, discharge current, pulse-on time, and pulse-off time with their respective levels chosen for the experimentation. The signal-to-noise ratios (S/N ratios) of the output responses were considered using the condition 'larger-is-better' for material removal rate and microhardness, whereas 'smaller-is-better' for surface roughness.

The experiments were carried out on the die-sinker-type EDM machine (model: S645 CMAX, make: OSCARMAX, Taiwan) selecting negative polarity conditions and a fixed machining time of 20 min for each run. Commercially available hydrocarbon oil, commonly known as EDM oil was used as a dielectric medium. Each trial was performed twice on two different plates to minimize the noise and errors, and further, an average of both was considered for the result analysis. Table 4 illustrates

**Table 4** Design of experiment matrix with their parametric combination

| Exp. run | Levels of process parameter | | | | Actual values of process parameter | | | |
|---|---|---|---|---|---|---|---|---|
| | A | B | C | D | Electrode | Current (Amp.) | *P*-on (μs) | *P*-off (μs) |
| 1 | 1 | 1 | 1 | 1 | Copper | 20 | 60 | 60 |
| 2 | 1 | 1 | 2 | 2 | Copper | 20 | 90 | 90 |
| 3 | 1 | 1 | 3 | 3 | Copper | 20 | 120 | 120 |
| 4 | 1 | 2 | 1 | 1 | Copper | 24 | 60 | 60 |
| 5 | 1 | 2 | 2 | 2 | Copper | 24 | 90 | 90 |
| 6 | 1 | 2 | 3 | 3 | Copper | 24 | 120 | 120 |
| 7 | 1 | 3 | 1 | 2 | Copper | 28 | 60 | 90 |
| 8 | 1 | 3 | 2 | 3 | Copper | 28 | 90 | 120 |
| 9 | 1 | 3 | 3 | 1 | Copper | 28 | 120 | 60 |
| 10 | 2 | 1 | 1 | 3 | Tungsten | 20 | 60 | 120 |
| 11 | 2 | 1 | 2 | 1 | Tungsten | 20 | 90 | 60 |
| 12 | 2 | 1 | 3 | 2 | Tungsten | 20 | 120 | 90 |
| 13 | 2 | 2 | 1 | 2 | Tungsten | 24 | 60 | 90 |
| 14 | 2 | 2 | 2 | 3 | Tungsten | 24 | 90 | 120 |
| 15 | 2 | 2 | 3 | 1 | Tungsten | 24 | 120 | 60 |
| 16 | 2 | 3 | 1 | 3 | Tungsten | 28 | 60 | 120 |
| 17 | 2 | 3 | 2 | 1 | Tungsten | 28 | 90 | 60 |
| 18 | 2 | 3 | 3 | 2 | Tungsten | 28 | 120 | 90 |

the experimental design matrix with their parametric combination based on the $L_{18}$ orthogonal array. The values for spark gap voltage (60 V), dielectric medium (EDM oil), and flushing pressure (0.5 kgf/cm$^2$) were kept constant throughout the experimentation.

For calculating the MRR, initial and final weight was measured for each run using a digital weighing balance (made citizen, model CY220), with display values up to three decimal places. Further, the material removal from the machined surface was calculated as per Eq. 1.

$$\text{MRR}\left(\text{mm}^3/\text{min}\right) = \frac{1000 \times \text{mass loss of workpiece}(g)}{\text{workpiece density}\left(\frac{g}{\text{mm}^3}\right) \times \text{machining time(min)}} \tag{1}$$

After the experimental process, the surface roughness (Ra, μm) of machined samples was measured using 'Mitutoyo surface roughness tester (SJ-401, Germany)' diametrically at three distinct points and arithmetic mean considered for the analysis. Moreover, the microhardness of the machined samples was examined using 'Mitutoyo microhardness tester (HM-220, Germany)' with a diamond indenter at an applied load of 0.98 N for a dwell time of 10 s. Three readings were taken at three different points, and an average was calculated for precise results. The selected output responses for the reported work were material removal rate ($mm^3$/min.), surface roughness (μm), and microhardness (HV).

## 3 Results and Discussion

Table 5 demonstrates the mean values of the output responses followed by a standard deviation (Avg. ± SD) and alongside the signal-to-noise (S/N) ratios for each trial. Relatively, the output responses were statistically investigated through analysis of variance (ANOVA) to scrutinize the significant factors affecting the particular output response, which are discussed in their respective sections.

**Table 5** Mean output response values and corresponding S/N ratio values

| Exp. run | Results | | | S/N ratios (dB) | | |
|---|---|---|---|---|---|---|
| | MRR ($mm^3$/min) Avg. ± SD | SR (μm) Avg. ± SD | MH (HV) Avg. ± SD | MRR | SR | MH |
| 1 | 2.714 ± 0.67 | 0.3894 ± 0.09 | 357.5 ± 6.23 | 51.0655 | 8.19208 | 8.6722 |
| 2 | 4.368 ± 0.61 | 0.4597 ± 0.08 | 385.6 ± 5.15 | 51.7227 | 6.75051 | 12.8057 |
| 3 | 6.071 ± 0.45 | 0.6282 ± 0.05 | 306.3 ± 7.42 | 49.7229 | 4.03804 | 15.6652 |
| 4 | 9.237 ± 0.13 | 0.4379 ± 0.04 | 371.8 ± 6.29 | 51.4062 | 7.17250 | 19.3106 |
| 5 | 10.615 ± 0.96 | 0.5673 ± 0.08 | 396.9 ± 6.14 | 51.9736 | 4.92374 | 20.5184 |
| 6 | 6.863 ± 0.29 | 0.9741 ± 0.06 | 495.1 ± 8.64 | 53.8939 | 0.22793 | 16.7303 |
| 7 | 10.292 ± 0.49 | 0.8756 ± 0.05 | 461.9 ± 8.49 | 53.2910 | 1.15388 | 20.2500 |
| 8 | 12.849 ± 0.21 | 0.6954 ± 0.04 | 429.4 ± 6.54 | 52.6572 | 3.15531 | 22.1774 |
| 9 | 10.411 ± 0.94 | 1.1083 ± 0.09 | 518.3 ± 9.28 | 54.2916 | −0.89315 | 20.3498 |
| 10 | 1.538 ± 0.98 | 0.4768 ± 0.08 | 389.7 ± 6.85 | 51.8146 | 6.43328 | 3.7391 |
| 11 | 2.0761 ± 0.37 | 0.7496 ± 0.05 | 514.2 ± 8.96 | 54.2226 | 2.50341 | 6.3450 |
| 12 | 2.4376 ± 0.30 | 0.6852 ± 0.06 | 372.9 ± 6.25 | 51.4318 | 3.28365 | 7.7392 |
| 13 | 3.6437 ± 0.14 | 0.8124 ± 0.06 | 506.5 ± 8.51 | 54.0916 | 1.80460 | 11.2309 |
| 14 | 4.2759 ± 0.66 | 0.7981 ± 0.02 | 463.1 ± 7.28 | 53.3135 | 1.95885 | 12.6206 |
| 15 | 3.9804 ± 0.47 | 1.0637 ± 0.07 | 531.8 ± 7.95 | 54.5150 | −0.53638 | 11.9985 |
| 16 | 5.2901 ± 0.84 | 0.9165 ± 0.08 | 594.9 ± 9.35 | 55.4889 | 0.75735 | 14.4693 |
| 17 | 7.2538 ± 0.66 | 1.1972 ± 0.07 | 678.5 ± 9.58 | 56.6310 | −1.56333 | 17.2113 |
| 18 | 6.7023 ± 0.34 | 0.8659 ± 0.08 | 651.6 ± 8.54 | 56.2796 | 1.25065 | 16.5245 |

## 3.1 Evaluation of Material Removal Rate

The material removal rate is a predominant response for the electric discharge machined substrate, which is majorly formulated by spark energy produced within the working area and the conductivity of the tool-workpiece during the process, i.e., more conductivity, more MRR [10–12]. On the same grounds, the sample machined (trial 8) with a copper electrode and higher current intensity (28A) depicts the superior MRR (12.849 $\pm$ 0.21 $mm^3$/min). Both current and electrode were the significant factors with a contribution of 55.58% and 33.92%, respectively. Table 6 listed the results of ANOVA, where significant factors were decided as per their $p$-value (<0.05). Figure 1 illustrates the S/N ratios plot for MRR generated using Minitab-17 software

**Table 6** Analysis of variance for signal-to-noise ratios of MRR

| Source | DF | Seq SS | Adj MS | $F$-value | $p$-value | % contribution |
|---|---|---|---|---|---|---|
| Electrode | 1 | 165.628 | 165.628 | 54.30 | 0.000* | 33.92 |
| Current | 2 | 271.373 | 135.686 | 44.48 | 0.000* | 55.58 |
| Pulse-on time | 2 | 18.433 | 9.217 | 3.02 | 0.094 | 3.77 |
| Pulse-off time | 2 | 2.366 | 1.183 | 0.39 | 0.688 | 0.48 |
| Residual error | 10 | 30.504 | 3.050 | | | 6.25 |
| Total | 17 | 488.303 | | | | 100 |

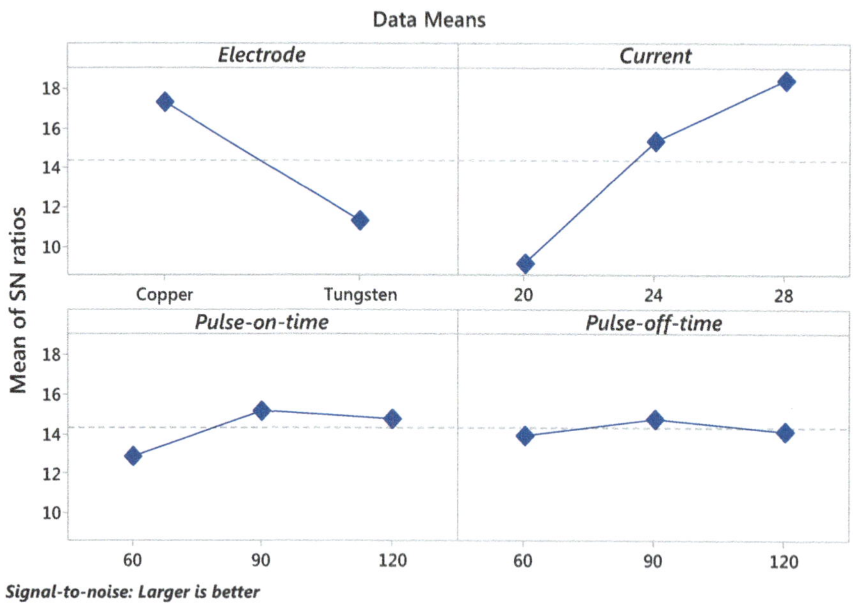


**Fig. 1** Main effects plot for S/N ratio of material removal rate

**Table 7** Analysis of variance for signal-to-noise ratios of SR

| Source | DF | Seq SS | Adj MS | *F*-value | *p*-value | % contribution |
|---|---|---|---|---|---|---|
| Electrode | 1 | 19.696 | 19.6957 | 5.91 | 0.035* | 13.60 |
| Current | 2 | 62.726 | 31.3631 | 9.42 | 0.005* | 43.29 |
| Pulse-on time | 2 | 27.614 | 13.8072 | 4.15 | 0.049* | 19.06 |
| Pulse-off time | 2 | 1.558 | 0.7788 | 0.23 | 0.796 | 1.08 |
| Residual error | 10 | 33.301 | 3.3301 | | | 22.97 |
| Total | 17 | 144.895 | | | | 100 |

following Taguchi's methodology for the 'larger-is-better' condition. The S/N ratios of MRR steeply increase as the intensity of current increases due to a rise in spark energy produced between the electrode and workpiece during the process [13, 14]. Under similar parametric settings, the copper electrode (trial 8) exhibits 77% more MRR compared to the sample machined with the tungsten electrode (trial 17). This improvement is due to the conductivity of the electrolytic copper tool, offering a better work function mechanism with the workpiece material.

### 3.2 Evaluation of Surface Roughness

Superior surface finish (0.3894 ± 0.09 μm) was examined for trial 1, machined with the copper electrode at lower current intensity 20 A, *P*-on 60 μs, *P*-off 60 μs. However, a drastic increase in SR was observed with tungsten electrode and the highest roughness exhibited by trial 17 (1.1972 ± 0.07 μm) at current 28 A, *P*-on 90 μs, *P*-off 60 μs. According to ANOVA results of Table 7, current (contribution: 43.29%), pulse-on duration (contribution: 19.06%), and electrode type (contribution: 13.60%) showed their dominance for the machining of SUS-316L under the selected parametric levels.

Similar results were revealed by Fig. 2 illustrating the effect of input process parameters on surface roughness. There is an inversely proportional relation of the surface finish with the spark energy produced on the working sample, i.e., the SR increases with an increase in the current and pulse-on duration. This is because of the explosion of intense energy in the working gap triggering the craters and micro-cracks on the machined surface and thereby increasing the roughness [15, 16].

### 3.3 Evaluation of Microhardness

The ANOVA results exhibiting the efficient factors associated with the microhardness of electric discharge machined SUS-316L. Superior microhardness of 678.5 ± 9.58 HV was revealed with the tungsten electrode at the utmost current of 28A

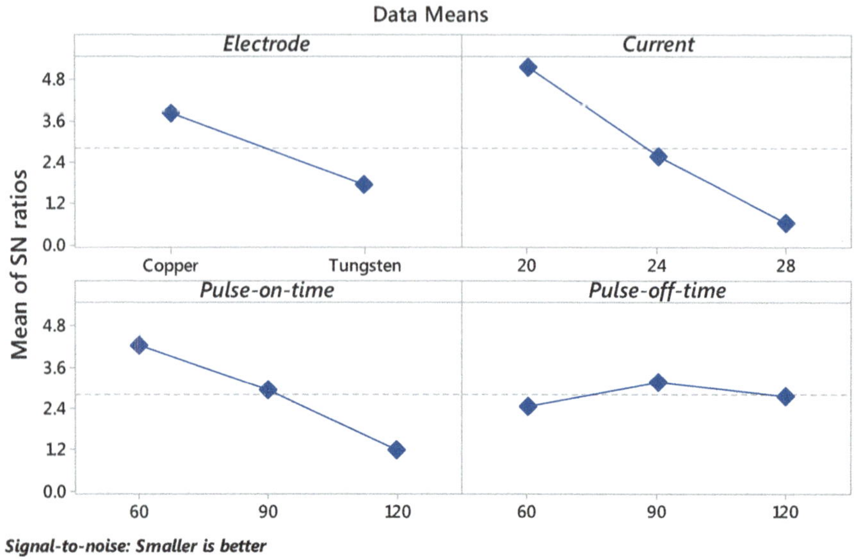


**Fig. 2** Main effects plot for S/N ratio of surface roughness

(trial 17) with an increment of 31% comparative to the sample machined with the same parametric setting using the copper electrode (trial 9). From the S/N ratios plot Fig. 3, the current and electrode are the dominating factors for the microhardness of the machined SUS-316L samples. The S/N ratios significantly increase with an increase in the current intensity coupled with the tungsten electrode. Similar results were portrayed by ANOVA Table 8, showing the major contribution of current (47.84%) and electrode (28.92%). It is due to the reason that high current produces the spark more rapidly, and because of the electrolyte (hydrocarbon oil) breakdown, intermetallic compounds formed and thus improved the surface hardness [17–19].

## 4 Conclusions

The present work reports the experimental comparison of electric discharge machined SUS-316L with copper and tungsten electrodes. From the results, it can be concluded that the current was the eminent factor for all three output responses. The surface machined with the copper electrode at parametric settings of current 28 A, *P*-on 90 μs, and *P*-off 120 μs exhibits MRR value 12.849 ± 0.21 $mm^3$/min, whereas the sample machined at lower levels of current (20 A), *P*-on (60 μs), *P*-off 60 μs with copper electrode showed the superior surface finish of 0.3894 ± 0.09 μm. The tungsten electrode significantly participated in the microhardness of the machined surface and revealed the higher microhardness (678.5 ± 9.58 HV) with an increase of 31% compared to the copper electrode under the same parametric values.

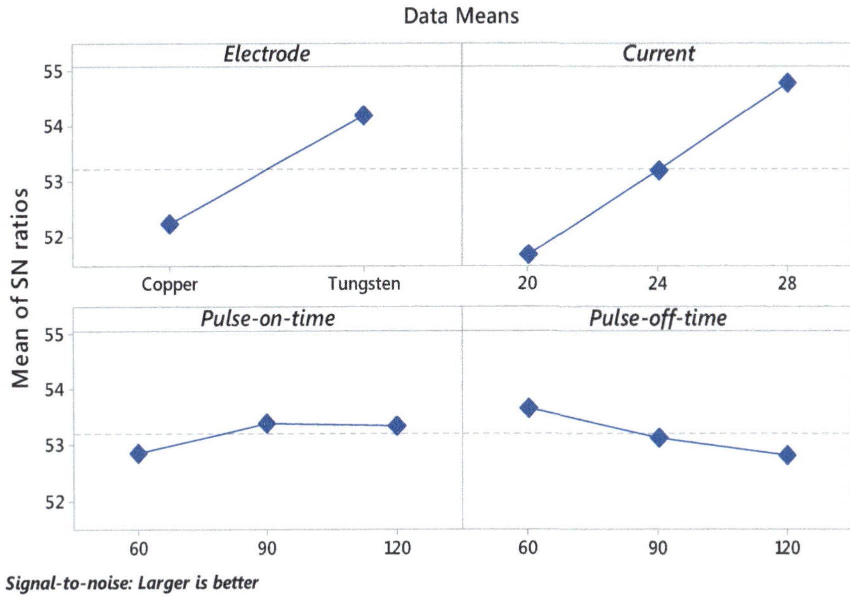


**Fig. 3** Main effects plot for S/N ratio of microhardness

**Table 8** Analysis of variance for signal-to-noise ratios of MH

| Source | DF | Seq SS | Adj MS | $F$-value | $p$-value | % contribution |
|---|---|---|---|---|---|---|
| Electrode | 1 | 17.531 | 17.5310 | 16.50 | 0.002* | 28.92 |
| Current | 2 | 29.015 | 14.5074 | 13.65 | 0.001* | 47.84 |
| Pulse-on time | 2 | 1.129 | 0.5645 | 0.53 | 0.604 | 1.86 |
| Pulse-off time | 2 | 2.347 | 1.1734 | 1.10 | 0.369 | 3.87 |
| Residual error | 10 | 10.625 | 1.0625 | | | 17.51 |
| Total | 17 | 60.647 | | | | 100 |

## References


1. Ho KH, Newman ST (2003) State of the art electrical discharge machining (EDM). Int J Mach Tools Manuf 43(13):1287–1300
2. Kumar S, Singh R, Singh TP, Sethi BL (2009) Surface modification by electrical discharge machining: a review. J Mater Process Technol 209(8):3675–3687
3. Singh G, Sidhu SS, Bains PS, Bhui AS (2019) Surface evaluation of ED machined 316L stainless steel in $TiO_2$ nano-powder mixed dielectric medium. Mater Today Proc 18(3):1297–1303
4. Srivastava V, Pandey PM (2013) Study of ultrasonic assisted cryogenically cooled EDM process using sintered (Cu–TiC) tooltip. J Manuf Process 15(1):158–166
5. Mahajan A, Sidhu SS, Devgan S (2020) MRR and surface morphological analysis of electrical-discharge-machined Co–Cr alloy. Emerg Mater Res 9(1):1–5

6. Bhui AS, Singh G, Sidhu SS, Bains PS (2018) Experimental investigation of optimal ED machining parameters for Ti-6Al-4V biomaterial. FU Series: Mech Eng 16(3):337–345
7. Guu YH, Hocheng H, Chou CY, Deng CS (2003) Effect of electrical discharge machining on surface characteristics and machining damage of AISI D2 tool steel. Mater Sci Eng A 358(1–2):37–43
8. Das MK, Kumar K, Barman TK, Sahoo P (2014) Application of artificial bee colony algorithm for optimization of MRR and surface roughness in EDM of EN31 tool steel. Procedia Mater Sci 6:741–751
9. Sharif S, Safiei W, Mansor AF, Isa MHM, Saad RM (2015) Experimental study of electrical discharge machine on stainless steel 316L using design of experiment. Procedia Manuf 2:147–152
10. Devgan S, Sidhu SS (2019) Evolution of surface modification trends in bone related biomaterials: a review. Mater Chem Phys 233:68–78
11. Mahajan A, Sidhu SS (2018) Surface modification of metallic biomaterials for enhanced functionality: a review. Mater Technol 33(2):93–105
12. Singh G, Sidhu SS, Bains PS, Singh M, Bhui AS (2020) On surface modification of Ti alloy by electro discharge coating using hydroxyapatite powder mixed dielectric with graphite tool. J Bio- Tribo-Corros 6(3):91
13. Pant P, Bharti PS (2020) Electrical Discharge Machining (EDM) of nickel-based nimonic alloys: a review. Mater Today: Proc 25:765–772
14. Mahajan A, Sidhu SS (2020) Devgan, S: Examination of hemocompatibility and corrosion resistance of electrical discharge-treated duplex stainless steel (DSS-2205) for biomedical applications. Appl Phys A 126:737
15. Al-Amin M, Abdul Rani AM, Abdu Aliyu AA, Abdul Razak MAH, Hastuty S, Bryant MG (2020) Powder mixed-EDM for potential biomedical applications: a critical review. Mater Manuf Process 1–23
16. Singh G, Sidhu SS, Bains PS, Bhui AS (2019) Improving microhardness and wear resistance of 316L by $TiO_2$ powder mixed electro-discharge treatment. Mater Res Exp 6(8):086501
17. Mahajan A, Sidhu SS (2019) Potential of electrical discharge treatment to enhance the in vitro cytocompatibility and tribological performance of Co–Cr implant. J Mater Res 34(16):2837–2847
18. Singh G, Ablyaz TR, Shlykov ES, Muratov KR, Bhui AS, Sidhu SS (2020) Enhancing corrosion and wear resistance of Ti6Al4V alloy using CNTs Mixed electro-discharge process. Micromachines 11(9):850
19. Devgan S, Sidhu SS (2020) Surface modification of β-type titanium with multi-walled CNTs/μ-HAp powder mixed electro discharge treatment process. Mater Chem Phys 239:122005

# Surface Integrity of Powder Mixed Electrical Discharge Treated Substrate at High Discharge Energies

**Sandeep Devgan, Amit Mahajan, Gurpreet Singh, Gurcharan Singh, and Sarabjeet Singh Sidhu**

**Abstract** The current study presents the investigation of powder mixed electrical discharge machining parameters in terms of surface integrity and machinability of β-type titanium alloy by utilizing three different dielectric medium. The three dielectric mediums, i.e., deionized water, multi-walled carbon nanotubes (MWCNTs) in deionized water and μHydroxyapatite powder (μHAp) in deionized water were employed for examination. To investigate the effectiveness of the powder mixed over the conventional electrical discharge process, the experiments were conducted according to the L18 experimental array, and metal removal rate (MRR) was studied. The results revealed that the machining in MWCNTs dielectric medium at 25 A current with negative polarity by incorporating copper/tungsten (Cu/W) electrode considered as optimal parameters of operation for higher material removal rate (MMR). MWCNTs dielectric medium significantly elevates the MRR to 58.7 $mm^3$/min. However, in the conventional electrical discharge machining, the feasibility of the process for achieving the good surface integrity was constrained at high current. Moreover, the powder mixed electrical discharge machining shifts the efficiency of machining. The best surface integrity was recognized with PM-EDM even at the higher ranges of discharge energies. The presence of powder particles in the dielectric medium generates a uniform-patterned surface without any surface irregularities.

**Keywords** Powder mixed electrical discharge machining · β-Titanium alloy · Carbon nanotubes · μHAp powder · Surface integrity

S. Devgan (✉) · A. Mahajan · G. Singh
Mechanical Engineering Department, Khalsa College of Engineering & Technology, Amritsar 143001, Punjab, India

G. Singh · S. S. Sidhu
Mechanical Engineering Department, Beant College of Engineering & Technology, Gurdaspur 143521, Punjab, India



P. Srinivasa Pai and V. Krishnaraj (eds.), *Sustainable Machining Strategies for Better Performance*, Lecture Notes in Mechanical Engineering,
https://doi.org/10.1007/978-981-16-2278-6_18

# 1 Introduction

Nowadays, non-conventional manufacturing processes are vastly recommended for the fabrication as well as for surface modification of complex shape components. The non-conventional fabrication methods provide effective cutting, accuracy and metal removal at a faster rate. However, the demand for a precise fabrication process is being increased due to speeding up in profound studies related to human anatomy or other complex-shaped joint replacements. The high precision and geometric accuracy are the major necessities for the effective functioning of implantations [1]. Electrical discharge machining (EDM) is a futuristic technique for fabricating complex profiles and attains a broad range of regular and textured surfaces by easily altering the process parameters [2, 3].

However, the conventional EDM process has some constraints related to manufacturing at high discharge energies. EDM-treated surfaces at high discharge energies exhibit surface irregularities like cracks and uneven blowholes due to a large temperature gradient during machining. The deprived surface quality leads to poor substrate precision and also reduces the service life of machined parts [4]. The EDM process parameters significantly affect the various aspects of surface integrity of the substrate as surface roughness elevates with the increase in peak current and pulse duration [5]. So, the suitable selection of process parameters for the best performance of EDM is a challenging task [6]. The powder mixed electrical discharge machining (PM-EDM) has emerged as one of the novel innovations for elevating the effectiveness of conventional EDM [7]. The powder addition in dielectric significantly affects the performance of machining. The presence of foreign particles in the spark gap affects the conductivity between the tool and workpiece that contributes to the proficiency of the machining process [8]. Like, Mohri et al. [9] studied the effect of silicon powder in a dielectric on the surface finish of H-13 die steel and concluded that the fine surface without any surface irregularities was attained. Sharma et al. [10] found the improvement in tribological and corrosion characteristics of titanium alloy during the hexagonal boron nitride powder mixed in deionized water. Similarly, Yu et al. [11] utilized gas-assisted holed electrodes for EDMing of titanium alloy and observed better MRR and surface integrity.

Recently, carbon nanotubes (CNTs) that emerge as dielectric medium powder have excellent characteristics which can shift the thermal and electrical properties of dielectric that can have positive effects on output parameters like surface integrity of machined surface [12]. Izman et al. [13] investigated the effects of multi-wall carbon nanotubes added into dielectric in the EDM process. They concluded that material removal rate and surface roughness are improved by about 7 and 9% relatively along with better surface characteristics of the modified surface. Similarly, the bioactive material like hydroxyapatite {(Ca10(PO4)6(OH)2, HAp} is potentially used as a powder in the dielectric medium in EDM. Basically, the HAp powder is a medical-grade material used to induce biocompatibility of alloy [14]. The HAp coatings are prolonged utilize to increase the bioactivity of metallic implants without compromising the bulk properties of an alloy [15]. Like, hydroxyapatite (HA) powder was utilized in PM-EDM to promote the biological performance and surface integrity of the substrate. Therefore, in the current article, three different dielectric mediums, i.e., deionized water, deionized water with multi-walled carbon nanotubes (MWCNTs) and deionized water with μHAp powder (micro-Hydroxyapatite powder) were used. The L18 Taguchi's experimental array was utilized to scrutinize the performance of PM-EDM process parameters on β-type titanium alloy. The comparison of the surfaces fabricated with three different dielectric medium was scrutinized in terms of surface integrity. Also, the most favorable set of process parameters was figure out for higher material removal rate (MRR).

## 2 Materials and Methods

The medical-grade β-type titanium alloy (Composition (Wt. %); Ti: 53%, Nb: 35%, Ta: 7%, Zr: 5%, O: 0.08%) was used for experimentation. The recent reported researches were contemplated as a resource to select the input parameters in the current experimentation [1, 3, 7]. The three different dielectric medium sets were used for investigation. The first medium was pure deionized water, and deionized water + μHAP powder with a concentration of 15 g/l was considered as the second medium. The third medium was the deionized water + MWCNTs with a concentration of 5 g/l. The three different materials, i.e., graphite (grained ~ 5 μm), tungsten and tungsten (75% wt) + copper (25% wt) (#NICKUNJ, New Delhi, India) were used as tool electrodes. For experimentation, the separate tank of capacity 50 L was fabricated along with a coupled pump (Flow Rate: 2000 lph; Max Head: 2.5 m) for effective flushing of dielectric medium (Fig. 1). The experimental trials are performed on EDM (Model: SZNC-35–5030) at assorted values of current and pulse on/off with

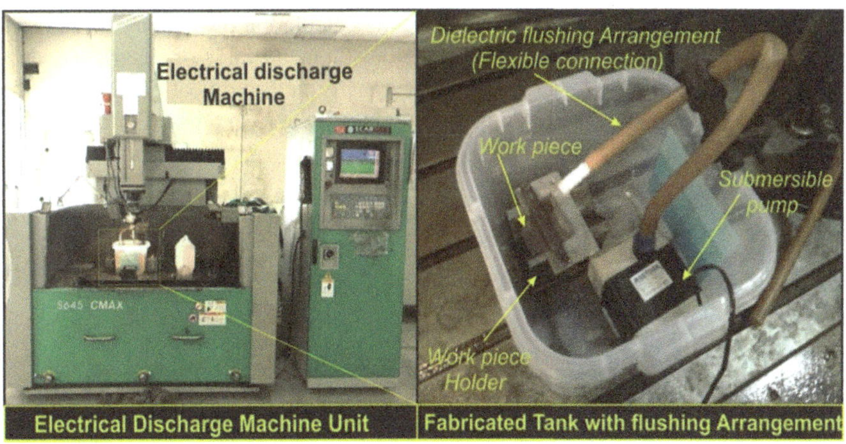


**Fig. 1** Schematic view of electrical discharge machining with dielectric flushing arrangement

**Table 1** Working factors with different levels

| Process parameters | Level |
|---|---|
| Polarities | Negative (N), positive (P) |
| Peak current ($I_P$)(Amp) | 10 Amp, 15 Amp, 25 Amp |
| Pulse-on ($P_{on}$)(μs) | 20 μs, 60 μs, 100 μs |
| pulse-off($P_{on}$)(μs) | 20 μs, 60 μs, 100 μs |
| Dielectric medium | Water, Water + HAp, Water + MWCNTs |
| Electrode material | Graphite (Gr), tungsten(W), copper–tungsten (Cu-W) |

both polarities (−/ + ) at predetermined working gap voltage between electrode and workpiece, i.e., 140 V.

## 2.1 Design of Experiment

Table 1 presents the input performance parameters and their respective levels. The Taguchi $L_{18}$ an orthogonal design of experimentation is illustrated in Table 2. The analysis of variance (ANOVA) was employed to examine the influence of input performance parameters on the signal-to-noise outcomes (S/N ratio) of the MRR by incorporating Minitab-17 statistical tool.

## 2.2 Discharge Energy Calculations

The discharge energies were scrutinized by the calculation of the duty cycle of all trials as described in Eq. 1. Discharge energy is the mean value of electrical energy produced per one impulse which is further transformed into heat energy and also represents as the product of peak current and duty cycle as expressed in Eq. 2.

$$\text{Duty Cycle}(\%) = \frac{\text{Pulse on Time}}{\text{Pulse on Time} + \text{Pulse off Time}} * 100 \tag{1}$$

$$\text{Discharge Energy} = \text{Duty Cycle} * \text{peak current} \tag{2}$$

**Table 2** Experimental array with MMR and EWR responses (Respective S/N Ratios)

| Process parameters | | | | | | | Responses | |
|---|---|---|---|---|---|---|---|---|
| Trials | Polarity (−/+) | $I_P$ (A) | $P_{on}$ (μs) | $P_{on}$ (μs) | Electrodes materials | Dielectric medium | MRR ($mm^3$/min) Avg ± SD | Discharge energy |
| 1 | N | 10 | 20 | 20 | Cu/W | D 1 | 11.457 ± 0.29 | 500 |
| 2 | N | 10 | 60 | 60 | Gr | D 2 | 2.423 ± 0.30 | 500 |
| 3 | N | 10 | 100 | 100 | W | D 3 | 42.353 ± 0.34 | 500 |
| 4 | N | 15 | 20 | 20 | Gr | D 2 | 3.433 ± 0.18 | 750 |
| 5 | N | 15 | 60 | 60 | W | D 3 | 43.387 ± 0.31 | 750 |
| 6 | N | 15 | 100 | 100 | Cu/W | D 1 | 31.583 ± 0.43 | 750 |
| 7 | N | 25 | 20 | 60 | W | D 1 | 33.413 ± 0.19 | 625 |
| 8 | N | 25 | 60 | 100 | Cu/W | D 2 | 30.733 ± 0.29 | 937.5 |
| 9 | N | 25 | 100 | 20 | Gr | D 3 | 58.773 ± 0.09 | 2083.3 |
| 10 | P | 10 | 20 | 100 | Gr | D 3 | 15.497 ± 0.25 | 166.6 |
| 11 | P | 10 | 60 | 20 | W | D 1 | 2.497 ± 0.13 | 750 |
| 12 | P | 10 | 100 | 60 | Cu/W | D 2 | 0.843 ± 0.11 | 625 |
| 13 | P | 15 | 20 | 60 | Cu/W | D 3 | 29.283 ± 0.24 | 375 |
| 14 | P | 15 | 60 | 100 | Gr | D 1 | 1.517 ± 0.39 | 562.5 |
| 15 | P | 15 | 100 | 20 | W | D 2 | 6.703 ± 0.19 | 1250 |
| 16 | P | 25 | 20 | 100 | W | D 2 | 8.243 ± 0.13 | 416 |
| 17 | P | 25 | 60 | 20 | Cu/W | D 3 | 50.843 ± 0.08 | 1875 |
| 18 | P | 25 | 100 | 60 | Gr | D 1 | 2.953 ± 0.13 | 1562.5 |

D 1: Deionized water
D 2: Deionized water with micro-hydroxyapatite powder of concentration 15 g/l
D 3: Deionized water with MWCNTs of concentration 5 g/l

## 2.3 *Material Removal Rate Calculations*

The material removal rate (MRR) was obtained by calculating the mass difference of each substrate (before and after the machining). The equation for evaluating the MRR in $mm^3$/min is:

$$(\text{MRR}) = \frac{1000 * \text{mass loss of workpiece(g)}}{\text{Density of worhpiece}\left(\frac{g}{mm^3}\right) * \text{time required for machining(min)}} \quad (3)$$

Moreover, the optimal machining parameters were also figured out for the best results of MRR using the statistical ANOVA technique. The best samples from MRR responses were further investigated in terms of surface integrity of substrates. Field emission scanning electron microscopy (FE-SEM) (JSM-7600F Jeol, Japan) was employed for the investigation of the surface integrity of substrates.

# 3 Results and Discussion

## 3.1 Statistical Analysis

Material removal rate (MRR) outcomes for all the trials are summarized in Table 2. The values listed are the means obtained from three attempts of each trial. The analysis of variance technique with the criteria 'the larger the better' at 95% confidence ($P < 0.05$) was employed for identifying the significant process parameters affecting MRR (Table 3). The factors that exhibited a larger F-value confirmed their significance in contributing to the metal removal rate (MRR). The results revealed that the dielectric (P-value; 0.0), polarity (P-value; 0.001), current (P-value; 0.003) and electrode (P-value; 0.024) were considered as a significant factor that majorly affects the MRR. Figure 2 reveals that the machining in MWCNTs dielectric medium at 25 A current with negative polarity by incorporating copper/tungsten (Cu/W) electrode considered as optimal parameters of operation for higher material removal rate (MMR). MWCNTs dielectric medium (Medium 1) significantly elevated MRR (more than 98%). Trial 9 exhibited the highest MRR (58.7 $mm^3$/min), and although, it was also machined at the highest discharge energy among all samples. This is due to fact that the high electrical and thermal conductivity of MWCNTs aid the electric spark to propagate more frequently into the heat-affected zone. Also, higher specific stiffness and structural toughness of MWCNTs facilitate the material exclusion process by bursting action.

The water-treated substrates also showed superior results of MRR as compared to μHAp mixed dielectric medium, indicated in Fig. 2. As the current increases, the material removal was enhanced in the case of deionized water but on the flip side, μHAp powder promulgated into molten metal and forms a dense coating on the substrate that leads to reduce the MRR. However, the selection of process polarity contributed significantly to improve the MRR. The negative polarity was considered as more effective than positive polarity for machining of hard material as titanium

**Table 3** Analysis of variance (ANOVA) for MMR

| Process parameters | DOF (f) | Adj. sum of sqrs. | Adj. mean of sqrs. | F-value | P-value |
|---|---|---|---|---|---|
| Polarity | 1 | 1076.07 | 1076.07 | 36.27 | 0.001 |
| Current | 2 | 1028.51 | 514 | 17.33 | 0.003 |
| P-on | 2 | 155.4 | 77.7 | 2.62 | 0.152 |
| P-off | 2 | 43.5 | 21.75 | 0.73 | 0.519 |
| Dielectric | 2 | 3376.66 | 1688.33 | 56.9 | 0.000 |
| Electrode | 2 | 441.88 | 220.94 | 7.45 | 0.024 |
| Error | 6 | 178.03 | 29.67 | | |
| Total | 17 | 6300.05 | | | |

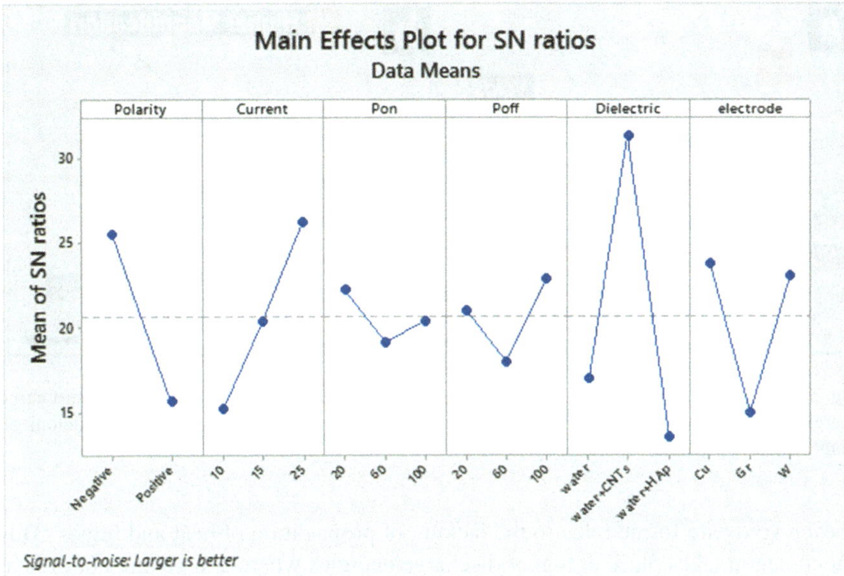


**Fig. 2** Main effects plot for S/N ratio of material removal rate

because higher momentum of positively charged ions produces bombardment on workpiece which causes more erosion action.

Similarly, the effect of electrode material was also significant for the material removal rate. The copper/tungsten electrode (Cu/W) was considered as the best electrode for metal removal of hard grade titanium alloy due to its high electrical and thermal conductivity which facilitates the heated electrons to immerse more rapidly in the molten bath and leads to less bulk electrical heating of electrode.

## 3.2 *Surface Morphological Analysis*

The calculated values of discharge energies were studied for the investigation of surface characteristics of machined surfaces. The trail 9 (MWCNTs-treated substrate) was machined at the highest discharge energy (i.e., 2083.3 J) among all the trials and also considered as a prominent sample in terms of MRR. Similarly, the trail 18 was machining at 1562.5 J discharge energy which was highest among all the trials treated with deionized water. In the case of μHAp treatment, trial 15 was machined at the highest discharge energy (i.e., 1250 J). Figure 3a, b represents the surface morphology of trial 15 (deionized water-treated substrate). The trail 18 exhibited a large amount of surface irregularities like voids, blowholes and uneven scattered debris of re-solidified molten metal (Fig. 3a). Moreover, the large number of surface micro-cracks was revealed at the high magnification (Fig. 3b). The blowholes and

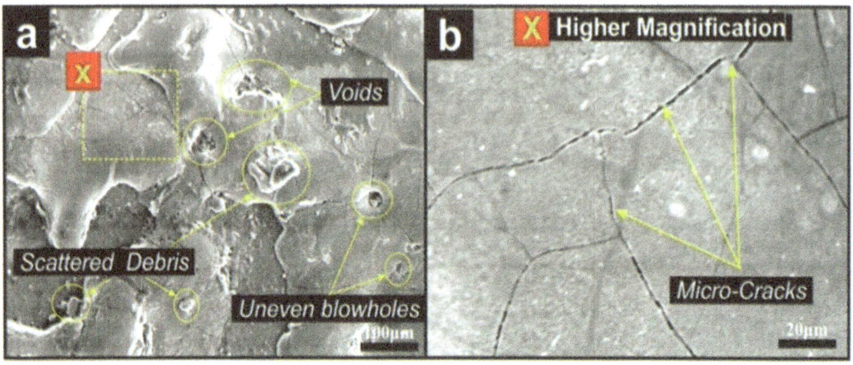


**Fig. 3** **a** Scanning electron microscopy of trail 18 and **b** magnified view of section X (discharge energy: 1562.5 J; peak current: 25 A; pulse on/off: 100 μs/60 μs; dielectric medium: deionized water; electrode: graphite, polarity: positive);

uneven voids are formed due to the lacking of propagation of heat and fumes. This phenomenon takes place at higher discharge energies where a large amount of heat is generated, and the plasma channel is unable to dissipate the heat flux from the workpiece. Due to insufficient propagation of heat flux, the large-sized voids and blowholes are shaped resulting in the formation of the non-uniform surface with high surface irregularities [16, 17].

Figure 4a, b demonstrates the surface morphology of powder mixed electrical discharge machining of trail 15 (μHAp treated substrate). Figure 4a reveals that the dense layer of μHAp powder was shaped onto the substrate and diminished the surface irregularities even at a higher value of discharge energy. This is due to the fact that the high specific density of μHAp particles tends to promulgate into molten metal and form dense coating onto the substrate that leads to resolving the cracks and other surface irregularities [18].

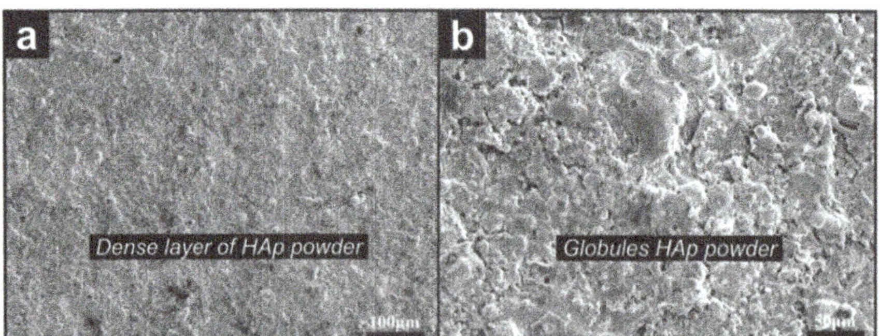


**Fig. 4** **a**, **b** Scanning electron microscopy of trail 15 and **b** magnified view of figure **a** (discharge energy: 1250 J; peak current: 15 A; pulse on/off: 100 μs/20 μs; dielectric medium: μHAp mixed deionized water; electrode: tungsten, polarity: positive)

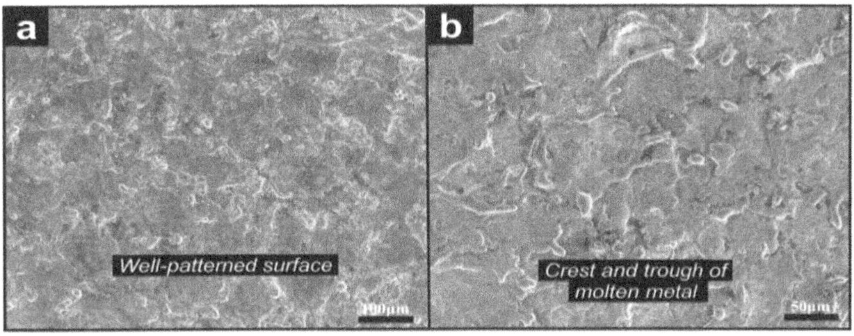


**Fig. 5** **a**, **b** Scanning electron microscopy of trail 9 and **b** magnified view of figure **a** (discharge energy: 2083.3 J; peak current: 25 A; pulse on/off: 100 μs/20 μs; dielectric medium: deionized water; electrode: graphite, polarity: negative)

Figure 5a, b illustrates surface micrographs of powder mixed electrical discharge machining of trail 9 (MWCNTs-treated substrate). Figure 5a, b represents the well-patterned structure without any surface irregularities. The MWCNTs facilitate to generate uniform surface and exhibit superior surface integrity. The presence of MWCNTs in the dielectric medium facilitates the smooth propagation of gathered energy by increasing the surface area plasma. The conductive MWCNTs have capabilities to form bridging action and shaped a well-spread heat flow net resulting in the formation of voids and craters free surfaces [19].

## 4 Conclusion

The investigation was laid on powder mixed electrical discharge machining (PM-EDM) in terms of material removal rate and the surface integrity of the medical-grade β-type titanium substrate by altering the process parameters. The study concluded that dielectric medium, process polarity, peak current and electrode material are the significant factor that majorly affects the metal removal rate (MRR). The machining in MWCNTs at 25 A peak current with negative polarity by incorporating copper/tungsten (Cu/W) electrode is considered as optimal parameters of operation for higher material removal rate (MMR). The MWCNTs mixed dielectric increased the MMR by 98% than deionized water due to its high conductivity. Thus, the powder mixed electrical discharge machining shifts the efficiency of machining to its prominent level. The well-patterned surfaces with the least surface irregularities were achieved in the case PM-EDM as compare to deionized water (conventional EDM) treatment. The best surface integrity was recognized with PM-EDM at the high discharge energy, whereas in the case of deionized water treatment, the feasibility of the process was constrained due to the formation of surface defects. In conclusion, the powder mixed electrical discharge machining (PM-EDM) is the prominent method to attain effective machinability along with desired surface integrity.

## References


1. Devgan S, Sidhu SS (2019) Evolution of surface modification trends in bone related biomaterials: a review. Mater Chem Phys 233:68–78
2. Mahajan A, Sidhu SS (2018) Surface modification of metallic biomaterials for enhanced functionality: a review. Mater Technol 33(2):93–105
3. Singh G, Sidhu SS, Bains PS, Singh M, Bhui AS (2020) On surface modification of ti alloy by electrodischarge coating using hydroxyapatite powder mixed dielectric with graphite tool. J Bio Tribo Corros 6:91
4. Al-Amin M, Abdul Rani AM, Abdu Aliyu AA, Abdul Razak MAH, Hastuty S, Bryant MG (2020) Powder mixed-EDM for potential biomedical applications: a critical review. Mater Manuf Process, pp 1–2
5. Pant P, Bharti PS (2020) Electrical discharge machining (EDM) of nickel-based nimonic alloys: a review. Mater Today: Proc 25:765–772
6. Mahajan A, Sidhu SS, Devgan S (2020) Examination of hemocompatibility and corrosion resistance of electrical discharge-treated duplex stainless steel (DSS-2205) for biomedical applications. Appl Phys A 126:737
7. Devgan S, Sidhu SS (2019) Enhancing tribological performance of β-titanium alloy using electrical discharge process. Surf Innov 8(1–2):115–126
8. Singh G, Sidhu SS, Bains PS, Bhui AS (2019) Improving microhardness and wear resistance of 316L by TiO2 powder mixed electro-discharge treatment. Mater Res Express 6(8):086501
9. Mohri N, Saito N, Higashi M, Kinoshita N (1991) A new process of finish machining on free surface by EDM methods. CIRP Ann 40(1):207–210
10. Sharma D, Mohanty S, Das AK (2020) Surface modification of titanium alloy using hBN powder mixed dielectric through micro-electric discharge machining. Surf Coat Technol 381:125–157
11. Yu YT, Hsieh SF, Lin MH, Huang JW, Ou SF (2020) Effects of gas-assisted perforated electrode with rotation on the machining efficiency of PMEDM of titanium. Int J Adv Manuf Technol 27:1
12. Chaudhury P, Samantaray S (2017) Role of carbon Nano Tubes in surface modification on electrical discharge machining-a review. Mater Today Proc 4(2):4079–4088
13. Izman S, Ghodsiyeh D, Hamed T, Rosliza R, Rezazadeh M (2012) Effects of adding multiwalled carbon nanotube into dielectric when EDMing titanium alloy. In: Advanced Materials Research. Trans Tech Publications Ltd, 463:1445–1449
14. Huang Y, Jin X, Zhang X, Sun H, Tu J, Tang T, Dai K (2009) In vitro and in vivo evaluation of akermanite bioceramics for bone regeneration. Biomaterials 30(28):5041–5048
15. Fang L, Leng Y, Gao P (2006) Processing and mechanical properties of HA/UHMWPE nanocomposites. Biomaterials 27(20):3701–3707
16. Sidhu SS, Bains PS (2019) Study of the recast layer of particulate reinforced metal matrix composites machined by EDM. Mater Today: Proc 4(2):3243–3251
17. Mahajan A, Sidhu SS (2019) Potential of electrical discharge treatment to enhance the in vitro cytocompatibility and tribological performance of Co–Cr implant. J. Mater. Res. 34(16):2837–2847

18. Devgan S, Sidhu SS (2020) Surface modification of β-type titanium with multi-walled CNTs/μ-HAp powder mixed electro discharge treatment process. Mater Chem Phys 239:122005
19. Singh G, Ablyaz TR, Shlykov ES, Muratov KR, Bhui AS, Sidhu SS (2020) Enhancing corrosion and wear resistance of ti6al4v alloy using cnts mixed electro-discharge process. Micromachines 11(9):850

# Analysis of Effect of Machining Parameters on Surface Roughness and MRR of AA3003/SiC Composite Material

**Sachinkumar Patil, M. Nagamadhu, K. Anand Babu, S. B. Kivade, and T. Veerbhadrappa**

**Abstract** In this research work, AA3003/SiC composite material is used as workpiece material for turning process using TNMG 16 04 08 insert. Experimentation was carried out using Taguchi $L_9$ orthogonal array. Three process parameters with three levels were selected. Chosen process parameters are speed (rpm), feed (mm/rev) and depth of cut (mm). Effect of these process parameters on surface roughness and material removal rate (MRR) was evaluated. ANOVA and signal to noise ratio analysis was carried out. ANOVA analysis showed that the speed is the major process parameters that affect the surface roughness and MRR severely compared to feed and depth of cut during turning process. Better surface finish and higher MRR was observed for speed of 1000 rpm, feed of 0.4 mm/rev and depth of cut of 2.5 mm. Better surface finish was observed for speed of 1200 rpm, feed of 0.6 mm/rev and depth of cut of 2 mm.

**Keywords** Aluminum alloys · Surface roughness · Composites · Machining · Material removal rate

## 1 Introduction

Today's manufacturing industries are in search of materials that possesses lightweight and better mechanical properties, this demand of automobile and aerospace industries is difficult to filled by the use of conventional alloys but can be fulfilled by the use of advanced materials such as composite materials [1]. The composite materials possesses attractive mechanical properties such as high strength to weight ratio, high

S. Patil (✉) · K. Anand Babu · T. Veerbhadrappa
School of Mechanical Engineering, REVA University, Bengaluru, India

M. Nagamadhu
Department of Mechanical Engineering, Acharya Institute of Technology, Bengaluru, India

S. B. Kivade
Department of Mechanical Engineering, Sri Jayachamarajendra College of Engineering, Mysuru, India



P. Srinivasa Pai and V. Krishnaraj (eds.), *Sustainable Machining Strategies for Better Performance*, Lecture Notes in Mechanical Engineering,
https://doi.org/10.1007/978-981-16-2278-6_19

modulus to weight ratio, higher wear resistance, higher coefficient of thermal expansion, higher corrosive resistance, excellent fatigue resistance, good damage tolerance, and therefore, these materials are becoming center of attraction for material research community [2]. As the research is going ahead in the field of composite materials, it is necessary to develop proper secondary manufacturing processes such as machining for these materials. Because machining is one of the important manufacturing process and machining cost plays major role in deciding the cost of a manufactured product. Machining cost can be reduced by optimizing the process parameters for machining a particular material. The demand for new cutting tools and economy in manufacturing all over the world will rise faster than industrial production as a whole in the near future. Turning process is one of the basic machining processes used to obtain cylindrical components. Usually, the machining process parameters were selected based on the skills and experience of the operators and also by referring the manual provided by machine tool manufacturer. This method may not be suitable for mass production where the higher precision and accuracy are necessary to meet customer demand. This is possible when there is a robust set of machining parameters is available for a particular material. Many of the researchers conducted studies on machining of materials by keeping one parameter constant and varying other parameter, which is a time consuming and not much accurate. Therefore, the use of optimization technique is helpful to get correct combination of process parameters with less time and less cost. There are different techniques are available for optimizing the process parameters among them Taguchi method is one of the most commonly used method [3]. Because it is simple and effective method. Taguchi statistical tool uses orthogonal array to optimize the process parameters. 3xxx series of aluminum alloys are used in many of the industrial applications, especially they are used in manufacturing of heat exchangers of power plants, packaging and chemical equipments. Especially, AA3003 is one of the widely used materials in 3xxx series of aluminum alloys due to their excellent formability properties and higher resistance to corrosion. But they possess medium strength; therefore, it is significant to incorporate the reinforcement particles such as SiC to base metals to enhance the strength. However, various literature are available to study the effect of machining parameters on different composites [4–6], but there is still lack of studies that need to be carried on machining of AA3003/SiC material. Hence, in the present work, AA3003/SiC composites were considered as material for turning. Accordingly, Taguchi tool is used to optimize the machining parameters during turning operation. The combined effect of process parameters on surface roughness and MRR is also studied. Experimental results are provided to confirm the effectiveness of Taguchi approach.

## 2 Experimentation

In the present work, cylindrical workpiece with 25 mm diameter AA3003/SiC material is considered for turning operation. Turning operation was carried out on a CNC machine. Various jobs were obtained for different combinations of process parameters. Liquid state processing such as stir casting route was employed for producing AA3003/SiC composites, because it is one of the simple and economical method for producing composites. For conducting the experiments according to Taguchi technique, three process parameters such as spindle speed, feed and depth of cut with three levels were considered. The selected levels are tabulated in Table 1. These levels were considered on the basis of **trial** experiments conducted. After turning process, surface roughness was measured using Talysurf as shown in Fig. 1. Material removal rate (MRR) was calculated and expressed as $mm^3$/min. In order to analyze the effect of machining process parameters, **namely**, spindle speed, feed and depth of cut and

**Table 1** Experimental result for MRR

| Trial no | Speed (rpm) | Feed (mm/rev) | DOC (mm) | MRR | S/N ratio | Mean |
|---|---|---|---|---|---|---|
| 1 | 800 | 0.2 | 1.5 | 4672 | 73.3901 | 4672 |
| 2 | 800 | 0.4 | 2 | 3876 | 71.7677 | 3876 |
| 3 | 800 | 0.6 | 2.5 | 4876 | 73.7613 | 4876 |
| 4 | 1000 | 0.2 | 2 | 5332 | 74.5378 | 5332 |
| 5 | 1000 | 0.4 | 2.5 | 3424 | 70.6907 | 3424 |
| 6 | 1000 | 0.6 | 1.5 | 2987 | 69.5047 | 2987 |
| 7 | 1200 | 0.2 | 2.5 | 3876 | 71.7677 | 3876 |
| 8 | 1200 | 0.4 | 1.5 | 2532 | 68.0693 | 2532 |
| 9 | 1200 | 0.6 | 2 | 4139 | 72.3379 | 4139 |

**Fig. 1** Talysurf used for surface roughness measurement

**Table 2** Experimental results for SR

| Trial no | Speed rpm) | Feed mm/rev) | DOC (mm) | Surface roughness (Ra) | S/N ratio | Mean |
|---|---|---|---|---|---|---|
| 1 | 800 | 0.2 | 1.5 | 22.23 | 26.9388 | 22.23 |
| 2 | 800 | 0.4 | 2 | 28.45 | 29.0816 | 28.45 |
| 3 | 800 | 0.6 | 2.5 | 19.65 | 25.8673 | 19.65 |
| 4 | 1000 | 0.2 | 2 | 25.78 | 28.2257 | 25.78 |
| 5 | 1000 | 0.4 | 2.5 | 31.54 | 29.9772 | 31.54 |
| 6 | 1000 | 0.6 | 1.5 | 24.33 | 27.7228 | 24.33 |
| 7 | 1200 | 0.2 | 2.5 | 27.23 | 28.7010 | 27.23 |
| 8 | 1200 | 0.4 | 1.5 | 22.12 | 26.8957 | 22.12 |
| 9 | 1200 | 0.6 | 2 | 19.23 | 25.6796 | 19.23 |

the interactions on the experimental data, analysis of variance is performed at 95% level. Before conducting the actual experiments, a set of trial and error experiments was conducted to find the range of cutting parameters for good surface finish and MRR. Tables 1 and 2 represent the ranking of each machining process parameter using Taguchi design of experiment (DOE) and analysis for S/N ratio, and means (larger is better) obtained at different process parameter levels.

## 3 Results and Discussion

Surface roughness is considered as one of the most important quality indicators during turning process. In the present work, surface roughness and MRR are considered as output responses. Surface roughness indicates the irregularities or unevenness present on the surface of manufactured parts. Process parameters such as cutting speed, feed and depth of cut affect the surface roughness and MRR. To obtain good surface finish and desired MRR, it is important to optimize the process parameter for a particular material. In this work, three process parameters with three levels were considered.

The S/N ratio for all machined components is determined by the following Eq. 1 (Sachinkumar et al. 2018).

$$\eta = -10\log\frac{1}{n}\sum_{i=1}^{n}\frac{1}{T_i^2} \tag{1}$$

In Eq. 1, $T_i$ is experimental value of the ith quality characteristic and n is the number of tests.

Followings are the steps involved in Taguchi parametric design optimization: (1) detection of process parameters to be optimized; (2) finding the number of levels for the selected parameters; (3) selecting orthogonal array; (4) running of experiments;

(5) analyzing the results using S/N ratio and ANOVA;(6) finding the optimal level of process parameters; (7) prediction of results; (8) running confirmation experiments.

Material removal rate (MRR) and surface roughness (SR) values were analyzed to study the effect of machining process parameters using Taguchi $L_9$ orthogonal array. Accordingly, the experimental values were calculated and tabulated in Tables 1 and 2, respectively, for MRR and SR. In this investigation, S/N ratio is selected according to the principle of "higher the better" in order to maximize the MRR. S/N ratio for SR selected according to the principle of "smaller the better" in order to minimize the SR. S/N ratios and means for MRR were calculated by using statistical software Minitab and presented in Tables 3 and 4, respectively. S/N ratio main plots are presented in Figs. 2a and 3a for MRR and SR, respectively. Means for MRR and SR are calculated and presented in Figs. 2b and 3b, respectively.

From Table 1 it can be noted that, higher MRR is observed for trial No. 4, i.e., for spindle speed of 1000 rpm, feed of 0.2 mm/rev and depth of cut of 2 mm, higher MRR is observed compared to other **trial** numbers. Amount of cutting force generated at this combination of process parameter may be sufficient that is why higher MRR is observed. But for 800 rpm and 1200 rpm, cutting force applied is not sufficient or excess as a result lower MRR is observed for these speeds, similar effect w.r.t. feed and depth of cut on MRR are observed.

Surface roughness values such as arithmetic mean (Ra) is measured for all machined components and also signal to noise ratios were calculated and tabulated in Table 2. Surface roughness was analyzed according to the principle of smaller the better. **Trial** No. 9 shows minimum surface roughness indicating better surface finish for the machined components obtained at spindle speed of 1200 rpm, feed of 0.6 mm/rev and depth of cut of 2 mm.

**Table 3** Response table for signal to noise ratios for MRR

| Level | Spindle speed (rpm) | Feed (mm/rev) | Depth of cut (mm) |
|---|---|---|---|
| 1 | 72.97 | 73.23 | 70.32 |
| 2 | 72.97 | 73.23 | 70.32 |
| 3 | 72.07 | 71.87 | 72.07 |
| Delta | 2.25 | 3.06 | 2.56 |
| Rank | 3 | 1 | 2 |

**Table 4** Response table for signal to noise ratios for SR

| Level | Spindle speed (rpm) | Feed (mm/rev) | Depth of cut (mm) |
|---|---|---|---|
| 1 | 27.30 | 27.96 | 27.19 |
| 2 | 28.64 | 28.65 | 27.66 |
| 3 | 27.09 | 26.42 | 28.18 |
| Delta | 1.55 | 2.23 | 1.00 |
| Rank | 2 | 1 | 3 |

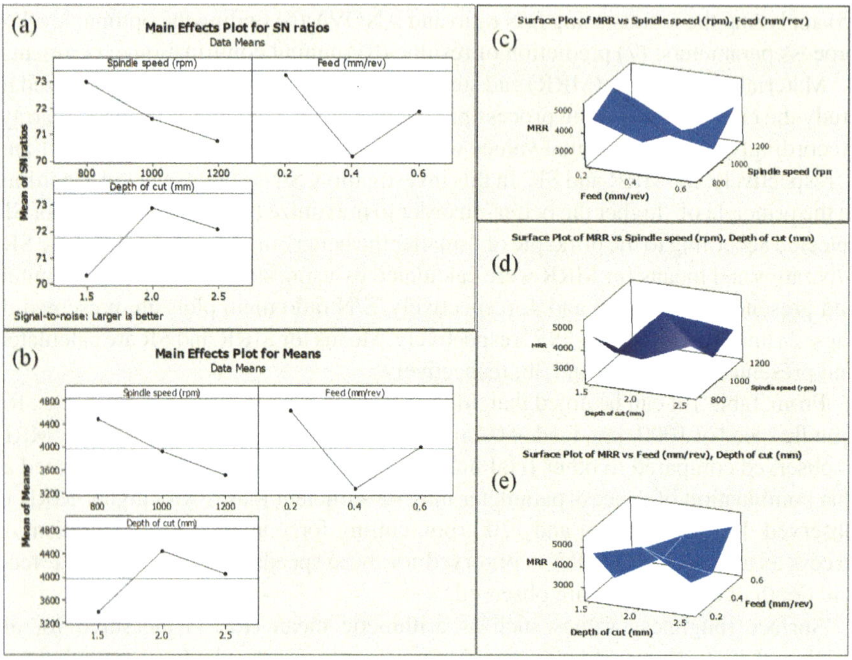


**Fig. 2.** **a** Main effects plots for S/N ratios for MRR, **b** main effects plots for means for MRR and surface plots for MRR showing combined effect of **c** spindle speed and feed **d** spindle speed and depth of cut, **e** feed and depth of cut.

ANOVA is one of the most widely used tools for analyzing the statistical data easily to identify the percentage contribution of each parameter on deciding the output response. In this work, ANOVA study was conducted at 95% confidence level. With the help of ANOVA study it is possible to get the clear pictures of effect of process parameters on output responses and the significance level of each process parameters. ANOVA results for MRR are presented in Table 5. From Table 5, it can be noted that feed rate plays major role in deciding MRR compared to speed and depth of cut. Therefore feed rate stands on rank 1, whereas depth of cut and speed stands at rank 2 and 3 respectively. Combined effect of each process parameters is analyzed through

**Table 5** ANOVA results for MRR

| Parameters | Degree of freedom | Sum of square | Mean square | F | Contribution rank |
|---|---|---|---|---|---|
| Spindle speed (rpm) | 2 | 1,392,590 | 696,295 | 0.403 | 3 |
| Feed (mm/rev) | 2 | 2,735,788 | 1,367,894 | 2.26 | 1 |
| Depth of cut (mm) | 2 | 1,696,867 | 848,433 | 1.60 | 2 |

surface plots and are presented in Fig. 2 (b, c and d) for MRR. Similarly, ANOVA results for SR are presented in Table 6. Feed is the major machining parameter affecting surface finish also, whereas depth of cut and speed stands at rank 2 and 3 respectively. Combined effect of each process parameters is analyzed through surface plots and are presented in Fig. 3 (b, c and d) for SR. Further, the theatrical results are confirmed by conducting confirmation test run and the results are presented in Table 7. From Table 7, it is clear that there is n much difference between the predicted values and experimental values, both the values are matched well.

**Table 6** ANOVA results for SR

| Parameters | Degree of freedom | Sum of square | Mean square | F | Contribution rank |
|---|---|---|---|---|---|
| Spindle speed (rpm) | 2 | 33.559 | 16.7794 | 1.60 | 2 |
| Feed (mm/rev) | 2 | 61.014 | 30.5071 | 2.92 | 1 |
| Depth of cut (mm) | 2 | 15.813 | 7.9065 | 0.53 | 3 |

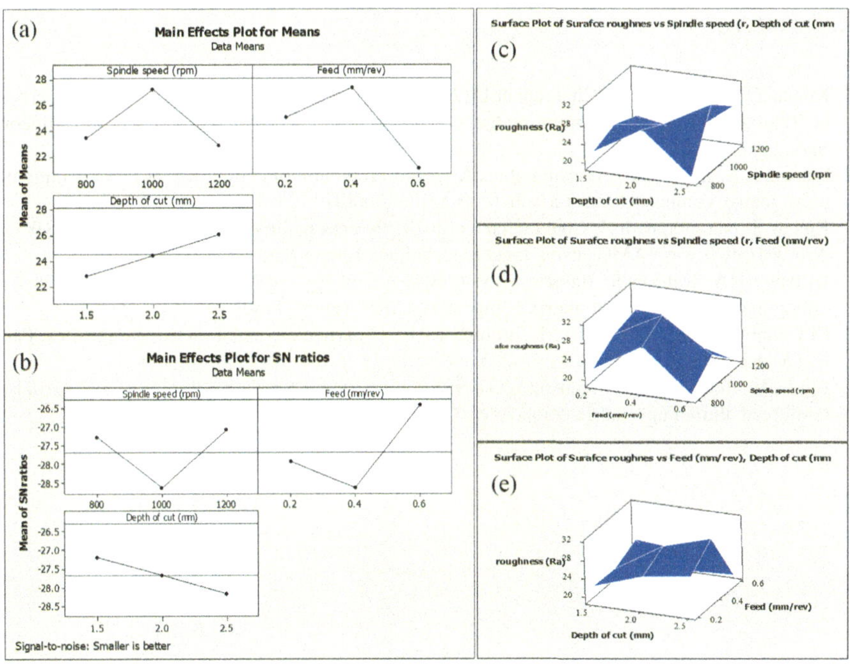


**Fig. 3** **a** Main effects plots for S/N ratios for SR, **b** main effects plots for means for SR and surface plots for SR showing combined effect of **c** spindle speed and feed **d** spindle speed and depth of cut, **e** feed and depth of cut

Table 7 Results of confirmation experiments

| Particulars | MRR | SR |
|---|---|---|
| Parameter level | A2B1C2 | A3B3C3 |
| Predicted value | 5220 | 19.68 |
| Experimental value | 5332 | 19.23 |

# 4 Conclusions

Turning process was used successfully for machining of AA3003/SiC composite material using TNMG 16 04 08 insert. Taguchi $L_9$ orthogonal array was applied and optimal combination of process parameters was identified to give best responses of MRR and SR. ANOVA analysis showed that the feed is the major process parameters that affect the surface roughness and MRR severely compared to speed and depth of cut during turning process for this material. Higher MRR was observed for speed of 1000 rpm, **feed of 0.2 mm/rev** and depth of cut of 2 mm. Better surface finish was observed for speed of 1200 rpm, feed of 0.6 mm/rev and depth of cut of 2 mm.

# References


1. Kumar S, Narendranath S, Chakradhar D (2018) Microstructure, hardness and tensile properties of friction stir welded aluminum matrix composite reinforced with SiC and Fly ash. Silicon Springer 11(6):2557–2565
2. Pramanik A, Zhang LC, Arsecularatne JA (2006) Prediction of cutting forces in machining of metal matrix composites. Int J Mach Tools Manuf 46(14):1795–1803
3. Kumar S, Narendranath S, Chakradhar D (2018) Process parameter optimization for FSW of AA6061/SiC/fly ash AMCs using taguchi technique. Emerg Mater Res 7(3):192–199
4. Kishawy HA, Kannan S, Balazinski M (2004) An energy based analytical force model for orthogonal cutting of metal matrix composites. CIRP Ann 53(1):91–94
5. El-Gallab M, Sklad M (1998) Machining of Al/SiC particulate metal matrix composites, Part II. J Mater Process Technol 84:277–285
6. Lin JT, Bhattacharyya D, Ferguson WG (1998) Chip formation in the machining of SiC-particle-reinforced aluminium-matrix composites. Compos Sci Technol 58(2):285–291

# Use of Vortex Tube Cooling for Machining Stellite 6

**G. Benaka, Bhaskara P. Achar, P. Srinivasa Pai, Grynal D'mello, and K. G. Gururaj**

**Abstract** Stellite alloys are extensively used for variety of applications in the modern industry. The machining of these alloys is of greater importance, considering the heat generated during machining. In addition to the difficulties in machining, the use of liquid coolants during the machining of these alloys causes a greater environmental impact. The manufacturing sector is continuously looking for better machining techniques for such alloys. This paper deals with the use of vortex tube cooling during the turning of Stellite 6 using uncoated carbide inserts. Various speeds (40, 60 and 80 m/min) were used in combination with various feeds (0.1, 0.15 and 0.2 mm/rev) for turning using cooled air emitted from the vortex tube. Tool tip temperature, surface roughness and tool flank wear were measured, and the results were compared with dry turning process. It was observed that the use of cooling air from the vortex tube played a significant part in decreasing the heat at the cutting zone and decreasing the flank wear. But the surface roughness was found to increase considerably compared to dry turning. Thus, vortex air cooling is a suitable alternative for machining of Stellite 6 alloys and is environmental friendly.

**Keywords** Vortex tube · Stellite 6 · Tool flank wear · Surface roughness · Air cooling · Tool tip temperature

## 1 Introduction

The applications of chromium-based alloys have increased rapidly in the past decade. Out of all the alloys of cobalt, there is significant growth in the usage of Stellite 6. Stellite, a trademarked name of Kennametal Inc., is a cobalt-chromium alloy which finds extensive usage in aerospace, automotive and chemical industries [1, 2]. Stellite 6 finds extensive usage due to its distinct properties. It possesses excellent strength, good wear resistance, resistance to creep and resistance to oxidation and corrosion [3,

G. Benaka · B. P. Achar (✉) · P. S. Pai · G. D'mello · K. G. Gururaj
Department of Mechanical Engineering, NMAM Institute of Technology, Nitte, 574110 Udupi, Karnataka, India
e-mail: bhaskarap@nitte.edu.in

 
P. Srinivasa Pai and V. Krishnaraj (eds.), *Sustainable Machining Strategies for Better Performance*, Lecture Notes in Mechanical Engineering,
https://doi.org/10.1007/978-981-16-2278-6_20

4]. On the contrary, machinability of these alloys is difficult. Due to this reason, they are termed as *"Difficult to cut materials"* [5]. The reason of difficulty in machinability is because of the high temperature in the machining zone. High temperature in the cutting zone is a consequence of low thermal conductivity [6]. Owing to the low thermal conductivity of these materials, the heat generated in the tool-work interface is not dissipated out fully. This concentration of heat in the machining zone is the major cause for accelerated tool wear and poor surface finish. In addition, turning of these alloys looks like an attractive option as a variety of parameters can be controlled during turning. But rapid work hardening of these alloys prevents the subsequent passes during machining. The most common solution to encounter these problems is the usage of liquid coolants. Ever since machining started, liquid coolants have been immensely popular. With all the advantages they carry, they also carry a greater disadvantage with their environmental impact [7]. Vortex tube is a simple mechanical device used to generate hot and cold streams of air from compressed air stream [4]. In present-day usage, the applications of vortex tube include cooling for machining applications, nuclear reactors, etc. Compressed air is taken in the inlet tangentially and separated into cold and hot streams of air from both the ends. The cold fraction of the vortex tube is the percentage of the air from the cold end [8]. The required cold fraction is a compromise between flow rate and temperature and depends on the application and requirement [9]. Higher the temperature through the cold air outlet, higher is the flow rate. The temperature and flow rate are controlled by the conical regulating valve. Zahari Taha et al. (2013) conducted studies on power consumed and surface roughness in dry turning using vortex tube cooling during turning of mild steel material using Tungaloy TNMG 160,408 TMT 9125 coated carbide inserts. The two conditions employed were ambient air and compressed air through vortex tube. Cutting speed of 100 m/min and 160 m/min, feed rates of 0.10, 0.18 and 0.28 mm/rev and a depth of cut of 1.0 mm to 4.0 mm were used to analyze the influence of tool wear and power consumed during machining. It was observed that better surface roughness was attained by ambient air cooling than vortex tube cooling due to the quenching of the surface being heat treated. Also surface finish was improved at lower feed rates than that of higher feed rates. Vortex tube application also demanded more power consumption due to the lack of lubrication during friction between the tool and the work piece [7]. H. Shao et al. (2013) conducted studies on turning of Stellite 12 alloy using SANDVIK coated SM1105 carbide tools and uncoated carbide tools YG 610 and YT726 in dry cutting conditions. Three cutting speeds and two feed rates with a constant depth of cut were used as the input parameters to evaluate the tool wear. Notch wear, maximum flank wear and average flank wear were observed. It was found that the material removal rate and the tool life of SM1105 were better than the uncoated carbide tools and also increased feed rates caused decrease in the tool life [5]. Mozammel Mia et al. (2018) conducted studies on the precision turning of Al 6061-T6 using CNMG 120,404 WIDIA tool inserts. The two cooling conditions involved were Ranque-Hilsch vortex tube and nitrogen gas-aided MQL. Two different cutting speeds and two different feeds were considered to evaluate the surface roughness parameter Ra. As cutting speed increased, the average surface roughness parameter Ra also increased [10]. This work deals with the turning of

Stellite 6 using uncoated carbide inserts under different cutting conditions at constant depth of cut using vortex tube. Tool tip temperature, surface roughness and tool flank wear were measured. The experiments were repeated in dry conditions, and the results were compared.

## 2 Experimental Details

The test specimen consisted of a cylindrical Stellite 6 bar of 60 mm diameter of 260 mm length. The chemical composition of Stellite 6 is given in Table 1 (as per the manufacturer).

The inserts used for turning were uncoated carbide with MR4 chip breaker (SECO), rhomboidal in shape, flat faced, nose radius of 0.8 mm with both the rake angles of -6° and end cutting edge angle of 5°. The tool holder used is PCLNL2020K12JETL (SECO). The tool overhang length was kept at 60 mm out of the total 120 mm length of the tool holder. The air cooling was achieved by using a small vortex tube #3208 (EXAIR). The compressed air required for the vortex tube was supplied by a 1 HP compressor (INTERNATIONAL COMPRESSOR) at 5.5–6.9 bar. The compressed air from the compressor was passed through a filter regulator 5612B262-B (SYNTESI), which separated moisture in the air by gravity. The vortex tube was rigidly mounted on one of the idle tool of the lathe turret. The controlling valve at the hot end of the vortex tube was adjusted, so that the cold fraction was set at 80%. This ensured maximum cooling with sufficient flow rate. The distance between the vortex tube cold air outlet and the tool tip was kept constant at 35 mm for all the experiments.

The vortex tube was kept in running condition for 30 s before the starting of the turning operation to stabilize the temperature and flow rate. The vortex tube setup is shown in Fig. 1. The conditions under which the experiments conducted are as mentioned in Table 2.

Each length of cut was taken as 40 mm. The output parameters measured are tool flank wear, surface roughness and tool tip temperature. The tool flank wear was measured in terms of maximum tool flank wear (VBmax). Tool maker's mjicroscope (Mitutoyo-TM505/510) was used to measure the flank wear after each pass with a magnification factor of 15 X. After each pass, the tool flank wear was recorded, and experiments were carried out till it reached 0.4 mm as per ISO 3685 [11]. Surface roughness was measured using a contact measuring device, Taylor and Hobson (Talysurf 50). Surface roughness was measured over a sample length of 2.5 mm. Value of surface roughness measured was the arithmetic average of surface profile (Ra). A

**Table 1** Chemical composition

| Element | Co | Cr | Mo | Fe | Si | Mn |
|---|---|---|---|---|---|---|
| Percentage | 63.5 | 27.8 | 6.6 | 0.9 | 0.8 | 0.4 |

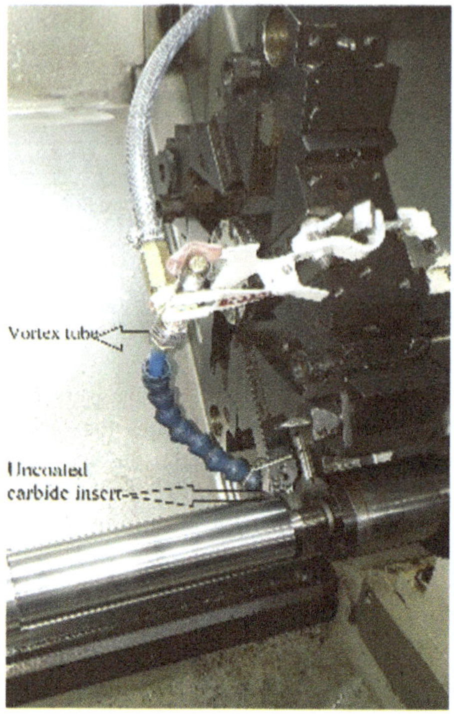


**Fig. 1** Vortex tube setup

**Table 2** Experimental conditions

| Parameters | Cutting speed (m/min) | Feed rate (mm/rev) | Depth of cut (mm) |
|---|---|---|---|
| Experimental conditions | 40 | 0.1 | 0.5 (Constant) |
| | 60 | 0.15 | |
| | 80 | 0.2 | |

30" dual laser infrared ray thermometer (EXTECH INSTRUMENTS) was used for measuring temperature at the cutting tool tip. The laser was focused on the tool-work interface and kept stable during the cut. The mean of four readings taken during the cut was calculated to find the average temperature.

## 3 Results and Discussion

The results discussed are for the experimental data obtained with vortex cooling compared with the data of dry machining of Stellite 6. The effect of different feeds and speeds was relatively compared.

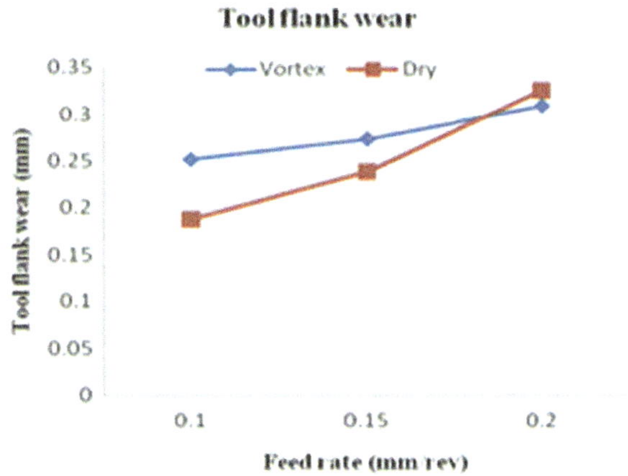


**Fig. 2** Influence of feed rate on tool flank wear at 80 m/min

In Fig. 2, the influence of different feed rates on tool flank wear at high-speed condition (80 m/min) is shown. It is observed that in both vortex cooled and dry machining condition, the tool flank wear rises with elevation in feed rate. This is owing to the high temperature and cutting forces involved during high cutting speed of 80 m/min [10]. Predominantly, the flank wear during dry turning is less than that of vortex cooled turning during lower feeds (0.1 mm/rev and 0.15 mm/rev). This is because at lower feed rate, the temperature reduction by vortex tube does not reduce the cutting forces involved. The wear is relatively higher due to the usage of uncoated carbide inserts [5]. Figure 3 shows the influence of feed rates on tool tip temperature at high-speed condition (80 m/min). It is witnessed that there is a considerable drop in the temperature of the tool tip during vortex cooled turning.

The maximum temperature recorded during vortex cooled turning was 52 °C when compared to 82 °C during the dry turning operation. This shows a decrease of 36.58% of temperature in the vortex cooled condition over the dry cutting condition.

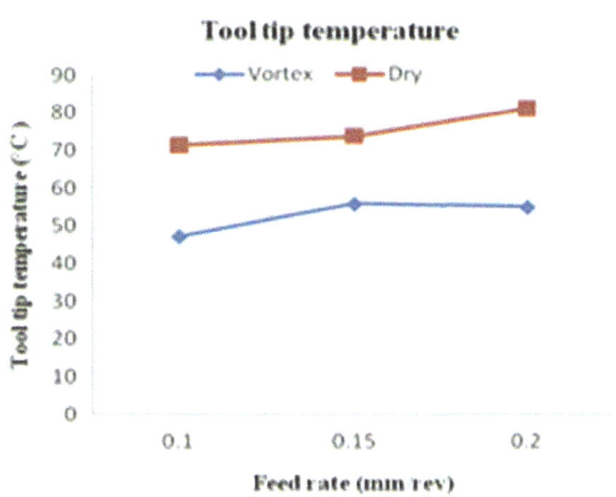


**Fig. 3** Effect of feed rate on tool tip temperature at 80 m/min

The increase in temperature in dry condition is attributed to the increased feed rate, which causes a growth in cutting forces and hence increases the temperature involved [5]. The decrease in temperature in vortex cooled condition is due to the cooling air of the vortex tube at the tool-work interface [12].

Figure 4 shows the effect of different feed rates on surface roughness at high-speed condition (80 m/min). It is observed that a decreasing trend is observed in the surface roughness during the dry turning, when compared to the increasing trend during the vortex cooled condition. This is because the cool air from the vortex tube rapidly cools the surface of the workpiece which has been softened due to the elevated temperature instantly [7]. Also, the low-temperature air causes the quenching of the surface that is being heat treated during machining [13]. This increases the surface roughness, when compared to dry turning where such conditions are not prevalent. Thus, the cooling air does not exhibit a chief role in improving the surface roughness [14]. Figure 5 shows the influence of different speeds on tool flank wear at high-feed

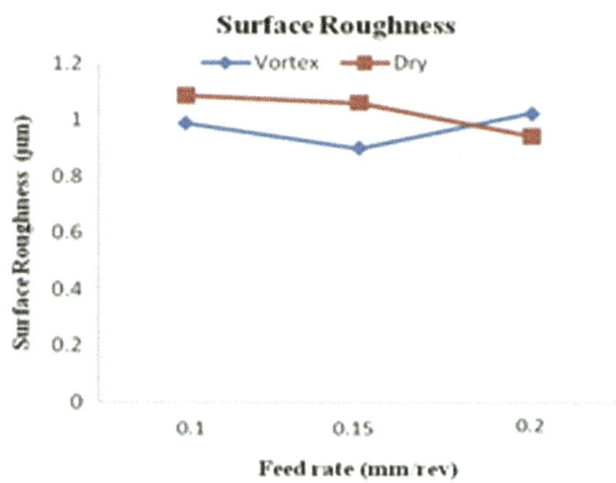


**Fig. 4** Influence of feed rate on surface roughness at 80 m/min

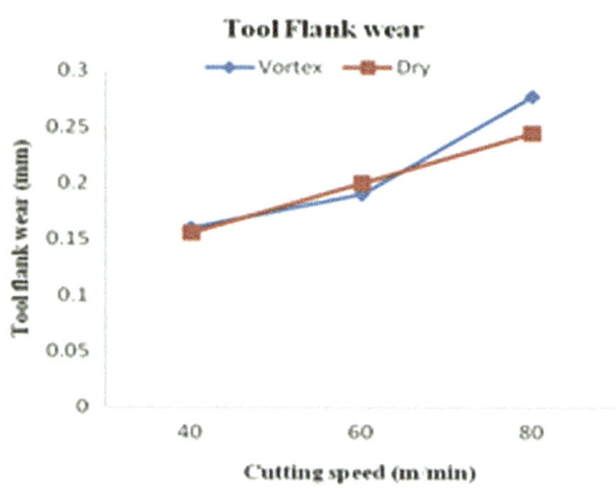


**Fig. 5** Influence of different speeds on flank wear at 0.2 mm/rev

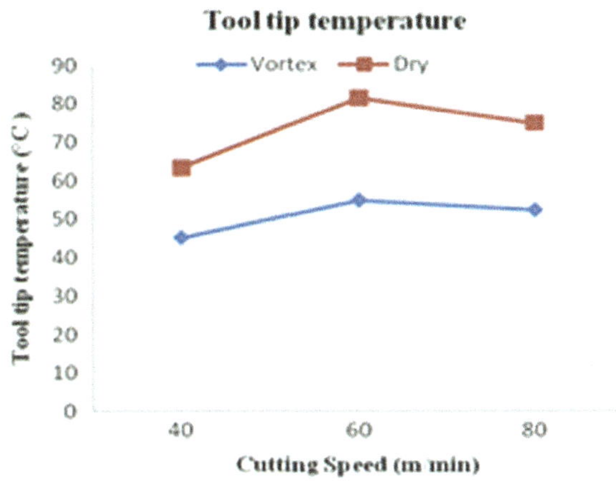


**Fig. 6** Effect of tool tip temperature with different speeds at 0.2 mm/rev

condition (0.2 mm/rev). It is seen that the tool flank wear in vortex cooled condition is lesser than dry cutting condition at lower speeds, except at high-speed condition (80 m/min). The tool flank wear is lesser in vortex cooled condition at low speed (40 m/min) and intermediate speed (60 m/min) due to the reduction in cutting forces which in turn prolongs the tool life [15]. The growth in tool flank wear at high speeds (80 m/min) is due to the increased temperature of the air in the cutting zone, which plays a negligible role in delaying the tool flank wear [15]. Thus, the cooling effect of air from vortex tube reduces at higher speed [16].

In Fig. 6, the influence of different speeds on tool flank wear at high-feed condition (0.2 mm/rev) is shown. It is found that there is a drastic reduction in the tool tip temperature in vortex cooled condition when compared to dry machining. The maximum temperature was recorded at 60 m/min and was 82 °C and 54 °C for dry and vortex cooled conditions, respectively. Thus, a reduction of 33.4% was observed in the vortex cooled condition. This is due to the reduction in temperature in the cutting zone due to the convection of heat through the cold air [17]. It is observed that the tool-chip temperature increased with the tool flank wear, due to the increase of friction between the tool and workpiece [16]. Figure 7 shows the effect of different speeds on surface roughness at high-feed condition (0.2 mm/rev). It is comprehended that the surface roughness increases in dry machining condition with increase in the cutting speed. At highest cutting speed condition of 80 m/min, the surface roughness of dry machining is more than that of vortex cooled condition. This is attributed to the increased temperature, which causes the micro-welding of the chip to the workpiece [10, 18]. Surface roughness is higher in vortex-cooled condition than dry condition at low speeds. This is due to cold air from the vortex tube, which caused difficulty in chip formation and shearing [17].

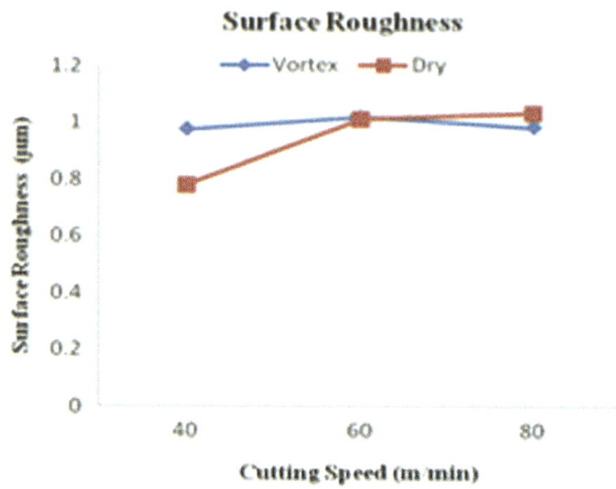


**Fig. 7** Effect of different speeds on surface roughness at 0.2 mm/rev

# 4 Conclusions

This study can be used as a reference for further work. Vortex-cooled air can be used as a suitable alternative for dry machining as well as use of flood coolant.

The present work deals with an eco-friendly way of machining Stellite 6, one of the most prominent alloys of the modern-day industry. It involves the use of Ranque-Hilsch vortex tube for the generation of cooling air. The following inferences can be made from the experimental investigations:

- The tool tip temperature decreases by 36.58 and 33.4% for maximum feed and maximum speed conditions, respectively, by the usage of cold air from the vortex tube.
- At maximum feed condition, the tool flank wear in case of vortex cooled condition is lesser than dry machining at all speeds (40 m/min and 60 m/min) except at maximum speed condition (80 m/min)
- At maximum speed condition, the tool flank wear in case of vortex cooled condition is lesser than dry machining at all feed rates (0.1 mm/rev and 0.15 mm/rev) except at maximum feed condition (0.2 mm/rev)
- The average surface roughness parameter Ra increased with the introduction of cooling air.

**Acknowledgements** The authors would like to thank N.M.A.M Institute of technology for the grants provided for carrying out the research work. The authors are grateful for the help and guidance extended by Dr.Sudesh Bekal, Dean R & D.

## References

1. Aykut Ş, Gölcü M, Semiz S, Ergür HS (2007) Modeling of cutting forces as function of cutting parameters for face milling of satellite 6 using an artificial neural network. J Mater Process Technol. 190:199–203
2. Agarwal SC, Ocken H (1990) The microstructure and galling wear of a laser-melted cobalt-base hardfacing alloy. Wear 140:223–233
3. Zaman HA, Sharif S, Kim DW, Idris MH, Suhaimi MA, Tumurkhuyag Z (2017) Machinability of Cobalt-based and Cobalt Chromium Molybdenum Alloys-A Review. Proc Manuf 11:563–570
4. Shokrani A, Dhokia V, Newman ST (2012) Environmentally conscious machining of difficult-to-machine materials with regard to cutting fluids. Int J Mach Tools Manuf 57:83–101
5. Shao H, Li L, Liu LJ, Zhang SZ (2013) Study on machinability of a stellite alloy with uncoated and coated carbide tools in turning. J Manuf Process 15:673–681
6. Safari H, Sharif S, Izman S (2014) Influence of cutting conditions on surface and sub-surface quality of high speed dry end milling Ti6Al-4V. J Teknol 3:131–135
7. Taha Z, Salaam HA, Ya TT, Phoon SY, Tan CF, Akiah MA (2013) Vortex tube air cooling: the effect on surface roughness and power consumption in dry turning. Int J Autom Mech Eng 8:1477
8. Inglis LR, Peter JE (1972) Vortec corp. Vortex tube cooling system. U.S. Patent 3,654,768
9. Singh G, Sharma VS (2017) Analyzing machining parameters for commercially puretitanium (Grade 2), cooled using minimum quantity lubrication assisted by a Ranque-Hilsch vortex tube. Int J Adv Manuf Technol 88:2921–2928
10. Mia M, Singh G, Gupta MK, Sharma VS (2018) Influence of Ranque-Hilsch vortex tube and nitrogen gas assisted MQL in precision turning of Al 6061-T6. Precision Engineering
11. Kuo C-P (2010) Sen-Chieh and Shao-Hsien Chen: tool life and surface integrity when milling Inconel 718 with coated cemented carbide tools. J Chin Inst Eng 33:915–922
12. Alsayyed B, Hamdan MO, Aldajah S (2012) Vortex tube impact on cooling milling machining. In: ASME 2012 international mechanical engineering congress and exposition, American Society of Mechanical Engineers, pp 773–776
13. Kalpakjian S, Schmid S (2010) Manufacturing, engineering and technology. SI 6th edn. Pearson, Delhi
14. Sharma VS, Dogra M, Suri NM (2009) Cooling techniques for improved productivity in turning. Int J Mach Tools Manuf 49:435–453
15. Yuan SM, Yan LT, Liu WD, Liu Q (2011) Effects of cooling air temperature on cryogenic machining of Ti–6Al–4V alloy. J Mater Process Technol 211:356–362
16. Sun S, Brandt M, Dargusch MS (2010) Machining Ti–6Al–4V alloy with cryogenic compressed air cooling. Int J Mach Tools Manuf 50:933–942
17. Su Y, He N, Li L, Iqbal A, Xiao MH, Xu S, Qiu BG (2007) Refrigerated cooling air cutting of difficult-to-cut materials. Int J Mach Tools Manuf 47:927–933
18. Raza SW, Pervaiz S, Deiab I (2014) Tool wear patterns when turning of titanium alloy using sustainable lubrication strategies. Int J Precis Eng Manuf 15:1979–1985

The manufacturer's authorised representative in the EU is Springer Nature Customer Service Centre GmbH, Europaplatz 3, 69115 Heidelberg, Germany. If you have any concerns regarding our products, please contact ProductSafety@springernature.com

Printed and bound by CPI Group (UK) Ltd, Croydon, CR0 4YY
30/09/2026
02225251-0003